#19

W9-BYL-768

Everyday Mathematics®

The University of Chicago School Mathematics Project

Skills Link

Courtney Tran

Cumulative Practice Sets
Student Book

Wright Group

The McGraw·Hill Companies

Photo Credits

Cover—©W. Perry Conway/CORBIS, *right;* Getty Images, *bottom left,* ©PIER/Getty Images, *center*

Photo Collage—Herman Adler Design

---

www.WrightGroup.com

 **Wright Group**

Copyright © 2009 by Wright Group/McGraw-Hill.

All rights reserved. Except as permitted under the United States Copyright Act, no part of this publication may be reproduced or distributed in any form or by any means, or stored in a database or retrieval system, without the prior written permission from the publisher, unless otherwise indicated.

Printed in the United States of America.

Send all inquiries to:
Wright Group/McGraw-Hill
P.O. Box 812960
Chicago, IL 60681

ISBN 978-0-07-6225057
MHID 0-07-6225054

3 4 5 6 7 8 9 MAL 15 14 13 12 11 10

Contents

Practice Sets Correlated to Grade 5 Goals

Content	*Everyday Mathematics* Grade 5 Grade-Level Goals	Grade 5 Practice Sets
Number and Numeration		
Place value and notation	**Goal 1.** Read and write whole numbers and decimals; identify places in such numbers and the values of the digits in those places; use expanded notation to represent whole numbers and decimals.	1, 2, 6, 7, 9, 12, 13, 16, 18, 23, 27, 34, 39, 41, 45, 47, 50, 51, 52, 56, 60, 61, 74, 75, 76
Meanings and uses of fractions	**Goal 2.** Solve problems involving percents and discounts; describe and explain strategies used; identify the unit whole in situations involving fractions.	12, 20, 28, 30, 31, 37, 42, 44, 59, 62, 63, 64, 75, 81, 87, 88
Number theory	**Goal 3.** Identify prime and composite numbers; factor numbers; find prime factorizations.	2, 3, 5, 6, 8, 10, 14, 16, 20, 21, 22, 23, 24, 28, 30, 31, 39, 45, 46, 47, 48, 49, 54, 64, 69, 76, 77, 80, 82, 85, 86
Equivalent names for whole numbers	**Goal 4.** Use numerical expressions involving one or more of the basic four arithmetic operations, grouping symbols, and exponents to give equivalent names for whole numbers; convert between base-10, exponential, and repeated-factor notations.	4, 6, 7, 8, 11, 40, 46, 48, 80, 85, 88
Equivalent names for fractions, decimals, and percents	**Goal 5.** Use numerical expressions to find and represent equivalent names for fractions, decimals, and percents; use and explain multiplication and division rules to find equivalent fractions and fractions in simplest form; convert between fractions and mixed numbers; convert between fractions, decimals, and percents.	5, 18, 28, 32, 33, 34, 35, 42, 45, 48, 49, 50, 54, 59, 62, 68, 71, 80, 82
Comparing and ordering numbers	**Goal 6.** Compare and order rational numbers; use area models, benchmark fractions, and analyses of numerators and denominators to compare and order fractions and mixed numbers; describe strategies used to compare fractions and mixed numbers.	1, 4, 5, 6, 9, 18, 20, 25, 30, 32, 35, 40, 48, 52, 53, 55, 56, 61, 67, 75, 79
Operations and Computation		
Addition and subtraction procedures	**Goal 1.** Use mental arithmetic, paper-and-pencil algorithms, and calculators to solve problems involving the addition and subtraction of whole numbers, decimals, and signed numbers; describe the strategies used and explain how they work.	2, 3, 4, 8, 9, 10, 11, 13, 15, 17, 24, 26, 30, 31, 38, 41, 42, 43, 51, 52, 53, 54, 64, 65, 71, 72
Multiplication and division facts	**Goal 2.** Demonstrate automaticity with multiplication facts and proficiency with division facts and fact extensions.	2, 3, 4, 6, 8, 10, 14, 15, 16, 20, 21, 22, 23, 24, 34, 40, 45, 48, 50, 76, 81, 82
Multiplication and division procedures	**Goal 3.** Use mental arithmetic, paper-and-pencil algorithms, and calculators to solve problems involving the multiplication of whole numbers and decimals and the division of multidigit whole numbers and decimals by whole numbers; express remainders as whole numbers or fractions as appropriate; describe the strategies used and explain how they work.	4, 6, 9, 19, 26, 27, 28, 31, 34, 37, 39, 42, 44, 46, 47, 50, 53, 54, 57, 60, 63, 71, 72, 78, 82, 89
Procedures for addition and subtraction of fractions	**Goal 4.** Use mental arithmetic, paper-and-pencil algorithms, and calculators to solve problems involving the addition and subtraction of fractions and mixed numbers; describe the strategies used and explain how they work.	43, 44, 51, 56, 57, 62, 67, 75, 83, 90
Procedures for multiplication and division of fractions	**Goal 5.** Use area models, mental arithmetic, paper-and-pencil algorithms, and calculators to solve problems involving the multiplication of fractions and mixed numbers; use diagrams, a common-denominator method, and calculators to solve problems involving the division of fractions; describe the strategies used.	48, 58, 59, 60, 65, 71, 80
Computational estimation	**Goal 6.** Make reasonable estimates for whole number and decimal addition, subtraction, multiplication, and division problems and fraction and mixed number addition and subtraction problems; explain how the estimates were obtained.	11,17, 25, 26, 67, 84
Models for the operations	**Goal 7.** Use repeated addition, arrays, area, and scaling to model multiplication and division; use ratios expressed as words, fractions, percents, and with colons; solve problems involving ratios of parts of a set to the whole set.	1, 6, 23, 24, 29, 30, 59, 63, 66, 88

Data and Chance

Data collection and representation	**Goal 1.** Collect and organize data or use given data to create bar, line, and circle graphs, with reasonable titles, labels, keys, and intervals.	39, 41, 42, 48, 51, 53, 74, 83, 90
Data analysis	**Goal 2.** Use the maximum, minimum, range, median, mode, and mean and graphs to ask and answer questions, draw conclusions, and make predictions.	12, 16, 36, 37, 38, 40, 41, 42, 52, 55, 67, 73, 76, 83, 84, 87
Qualitative probability	**Goal 3.** Describe events using *certain, very likely, likely, unlikely, very unlikely, impossible* and other basic probability terms; use *more likely, equally likely, same chance, 50–50, less likely,* and other basic probability terms to compare events; explain the choice of language.	8, 12, 16, 47, 86, 88
Quantitative probability	**Goal 4.** Predict the outcomes of experiments, test the predictions using manipulatives, and summarize the results; compare predictions based on theoretical probability with experimental results; use summaries and comparisons to predict future events; express the probability of an event as a fraction, decimal, or percent.	12, 18, 86, 90

Measurement and Reference Frames

Length, weight, and angles	**Goal 1.** Estimate length with and without tools; measure length with tools to the nearest $\frac{1}{8}$ inch and millimeter; estimate the measure of angles with and without tools; use tools to draw angles with given measures.	10, 30, 39, 53, 64
Area, perimeter, volume, and capacity	**Goal 2.** Describe and use strategies to find the perimeter of polygons and the area of circles; choose and use appropriate formulas to calculate the areas of rectangles, parallelograms, and triangles, and the volume of a prism; define *pi* as the ratio of a circle's circumference to its diameter.	5, 9, 23, 33, 35, 43, 58, 66, 68, 69, 70, 77, 78, 81, 82, 83, 84, 86, 88, 89
Units and systems of measurement	**Goal 3.** Describe relationships among U.S. customary units of length; among metric units of length; and among metric units of length; and among U.S. customary units of capacity.	14, 38, 40, 70, 83
Coordinate systems	**Goal 4.** Use ordered pairs of numbers to name, locate, and plot points in all four quadrants of a coordinate grid.	2, 34, 64, 65, 66, 75, 83

Geometry

Lines and angles	**Goal 1.** Identify, describe, compare, name, and draw right, acute, obtuse, straight, and reflex angles; determine angle measures in vertical and supplementary angles and by applying properties of sums of angle measures in triangles and quadrangles.	17, 18, 21, 22, 27, 50, 56, 80
Plane and solid figures	**Goal 2.** Describe, compare, and classify plane and solid figures using appropriate geometric terms; identify congruent figures and describe their properties.	11, 19, 20, 22, 23, 40, 58, 79, 80
Transformations and symmetry	**Goal 3.** Identify, describe, and sketch examples of reflections, translations, and rotations.	22, 58

Patterns, Functions, and Algebra

Patterns and functions	**Goal 1.** Extend, describe, and create numeric patterns; describe rules for patterns and use them to solve problems; write rules for functions involving the four basic arithmetic operations; represent functions using words, symbols, tables, and graphs and use those representations to solve problems.	1, 9, 6, 7, 14, 34, 36, 41, 47, 53, 57, 60, 62, 65, 67, 69 73, 76, 81, 86
Algebraic notation and solving number sentences	**Goal 2.** Determine whether number sentences are true or false; solve open number sentences and explain the solutions; use a letter variable to write an open sentence to model a number story; use a pan-balance model to solve linear equations in one unknown.	7, 18, 25, 30, 36, 49, 50, 51, 56, 57, 60 ,62, 71, 72, 73, 75, 87
Order of operations	**Goal 3.** Evaluate numeric expressions containing grouping symbols and nested grouping symbols; insert grouping symbols and nested grouping symbols to make number sentences true; describe and use the precedence of multiplication and division over addition and subtraction.	2, 4, 7, 9, 14, 16, 18, 27, 32, 44, 45, 52, 56, 57, 64, 82, 84
Properties of the arithmetic operations	**Goal 4.** Describe and apply the properties of arithmetic.	9, 25, 33, 57

Grade 4 Review: Number and Numeration

Write these decimals in standard form.

1. twenty three hundredths _____

2. three and eight hundred fifty two thousandths _____

3. fourteen and nine thousandths _____

4. one hundred sixty two and seventy hundredths _____

5. Four friends share 32 tokens at the arcade. If they share them equally, what fraction of the tokens will each person get? _____

How many tokens will each person get? _____

6. Write the next 5 multiples of 7.

7, _____, _____, _____, _____, _____

7. Write all the factors of 36.

1, _____, _____, _____, _____, _____, _____, _____, 36

Make a name-collection box for each number. Use as many different operations as you can. Use parentheses in at least two expressions.

8.

9.

Write the fraction as a decimal and a percent.

10. $\frac{3}{4}$ _____

11. $\frac{4}{5}$ _____

12. $\frac{7}{10}$ _____

13. $\frac{6}{100}$ _____

Order the numbers from least to greatest.

14. $\frac{2}{3}, \frac{5}{8}, \frac{3}{4}, \frac{1}{6}, \frac{7}{12}$

15. 0.94, 9.40, 0.094, 94.0, 9.04

_____, _____, _____, _____, _____   _____, _____, _____, _____, _____

4

Grade 4 Review: Data and Chance

1. Use the data in the table below to create a line graph showing the attendance at Westbrook Middle School's football games. Label each axis and give the graph a title.

Game	1	2	3	4	5	6	7	8	9	10	11	12
Attendance	180	220	165	240	205	190	200	205	240	160	175	205

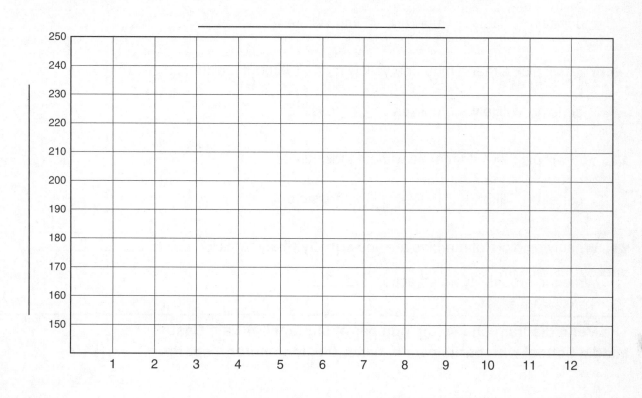

2. Which games had the maximum attendance? _____

3. Which game had the minimum attendance? _____

4. What is the range of attendance for all 12 games? _____

5. What is the median attendance for all 12 games? _____

6. What is the mode? _____

7. What is the mean? _____

Grade 4 Review: Data and Chance

**There are 3 red marbles, 8 blue marbles, 3 green marbles,
4 yellow marbles, and 2 orange marbles in a bag. Write the
fraction for the probability of each event. Then circle
the word that best describes the likelihood of that event.**

8. You will pick an orange marble without looking. _____

     certain   likely   unlikely   impossible

9. You will pick either a blue or a yellow marble without looking. _____

     certain   likely   unlikely   impossible

10. You will pick a white marble without looking. _____

     certain   likely   unlikely   impossible

11. What two colors of marbles are you **equally likely** to pick? _____

    Write the probability as a fraction. _____

**The Venn diagram shows the number of students at East Castle
Middle School who participate in band, theater, and/or a sport.**

12. How many students participate in
band and/or theater but do not play a sport? _____

13. How many total students participate in a sport? _____

14. How many students participate in
any two, but not three, of the activities? _____

15. How many students participate in all three activities? _____

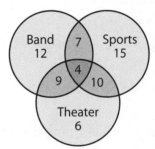

8

Grade 4 Review: Measurement and Reference Frames

Use a ruler to find the measurements.

1. Measure the marker to the nearest $\frac{1}{2}$ centimeter. _____ cm

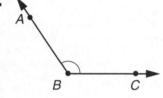

2. Measure the base of the stapler to the nearest $\frac{1}{4}$ inch. _____ in.

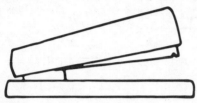

Tell whether the angle is *acute, right,* or *obtuse*. Then circle the best estimate of the measure of the angle.

3.

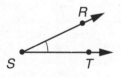

_____

  80°   95°   125°   170°

4.

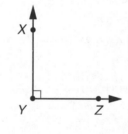

_____

  24°   80°   90°   180°

5.

_____

  10°   25°   90°   250°

6.

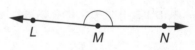

_____

  100°   175°   180°   190°

Grade 4 Review: Measurement and Reference Frames

7. Find the perimeter and the area of the shape.

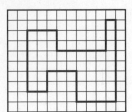

Perimeter: _____   Area: _____

8. Estimate the area of the shape.

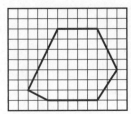

Area: about _____ square units

9.

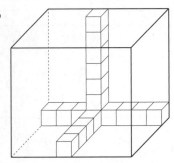

Volume = _____ cubes

10.

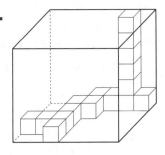

Volume = _____ cubes

Complete.

11. 164 cm = _____ m

12. 72 cm = _____ mm

13. 15 ft 3 in. = _____ in.

14. 1760 yd = _____ ft

Plot and label each point on the coordinate grid.

15. *A* (6,3)

16. *B* (2,9)

17. *C* (0,7)

18. *D* (9,1)

19. *E* (3,6)

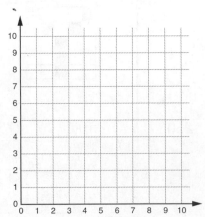

10

Grade 4 Review: Geometry

Use the grid to complete Problems 1–3.

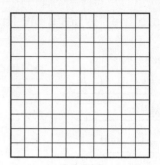

1. Draw and label a line *AB*.

2. Draw a line segment *CD* parallel to line *AB*.

3. Draw a ray *YZ* that intersects both lines *AB* and *CD*.

---

4. Draw 2 examples of shapes that are polygons.

5. Draw 2 examples of shapes that are **not** polygons.

Use the figure to answer the questions.

6. Name the figure. _____

7. How many faces does the figure have? _____

8. How many edges does the figure have? _____

9. How many vertices does the figure have? _____

---

10. Shade the reflection of the shape.

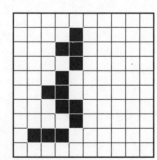

11. Shade what the figure will look like after it is rotated $\frac{1}{2}$ turn.

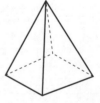

12. Translate the figure 4 spaces to the right.

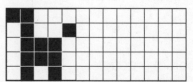

Name	Date	Time

Grade 4 Review: Patterns, Functions, and Algebra

Complete the "What's My Rule?" tables and state the rules.

1.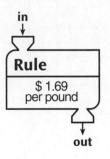

pounds	nuts
3	
5	
7	
9	
11	

2.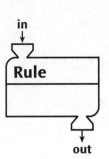

in	out
14	28
19	
21	42
	74
50	

---

3. Solve the open number sentence.

$64 = a \times (25 - 17)$ $a =$ _____

4. Jamal is baking cookies to sell. He sells the cookies for $3.75 a dozen. He wants to earn $45.00 to buy a new skateboard. Write and solve an open number sentence to show how many dozens of cookies he will need to sell.   _____

Solve.

5. $7 + (6 * 2) =$ _____

6. $56 - (12 + 17) =$ _____

Insert parentheses to make the equation true.

7. $12 + 6 * 3 = 54$

8. $25 \div 7 - 2 = 5$

Calculate the partial products. Then find the sum.

9. $14 * 27 = (10 + $ _____$) * ($ _____$ + 7)$

$10 *$ _____ $=$ _____

$10 * 7 =$ _____

$4 *$ _____ $=$ _____

_____ $* 7 =$ _____

Sum of partial products: _____

Practice Set 1

SRB
10 57
249

Write the letter of the array that matches the number model.

1. 2 * 5 = 10 _____   **A.** ● ● ● ● ● ● ● ●

2. 4 * 3 = 12 _____   **B.** ● ● ● ●
                                                    ● ● ● ●

3. 1 * 8 = 8 _____   **C.** ● ● ● ●
                                                    ● ● ● ●
                                                    ● ● ● ●

4. 3 * 4 = 12 _____   **D.** ● ● ● ● ●
                                                    ● ● ● ● ●

5. 2 * 4 = 8 _____   **E.** ● ● ●
                                                    ● ● ●
                                                    ● ● ●
                                                    ● ● ●

Round to the nearest hundred.

6. 699 _____   **7.** 2,208 _____   **8.** 7,942 _____

9. 51,985 _____   **10.** 22,761 _____   **11.** 17,032 _____

Complete the following number lines.

12.

0 _____ 0.4 _____ _____ 1

13.

15 _____ 25 30 _____

14.

4 _____ _____ 16 _____ 24

15.

_____ _____ _____ $\frac{3}{5}$ _____ 1

Practice Set 2

 FACTS PRACTICE For each set of problems, do as many problems as you can in one minute. You can ask someone to time you.

Problem Set 1	Problem Set 2	Problems Set 3
1. 5 * 2 = 10	**16.** 9 * 4 = 36	**31.** 2 * 11 = 22
2. 6 * 10 = 60	**17.** 7 * 8 = 56	**32.** 3 * 10 = 30
3. 2 * 4 = 8	**18.** 8 * 3 = 24	**33.** 1 * 4 = 4
4. 3 * 5 = 15	**19.** 9 * 2 = 18	**34.** 4 * 3 = 12
5. 8 * 6 = 48	**20.** 9 * 6 = 54	**35.** 3 * 7 = 21
6. 8 * 7 = 56	**21.** 5 * 10 = 50	**36.** 5 * 3 = 15
7. 11 * 5 = 55	**22.** 5 * 6 = 30	**37.** 11 * 0 = 0
8. 7 * 6 = 42	**23.** 4 * 8 = 32	**38.** 9 * 5 = 45
9. 5 * 4 = 20	**24.** 6 * 3 = 18	**39.** 5 * 8 = 40
10. 6 * 7 = 42	**25.** 3 * 8 = 24	**40.** 6 * 2 = 12
11. 7 * 9 = 63	**26.** 10 * 10 = 100	**41.** 4 * 4 = 16
12. 7 * 4 = 28	**27.** 9 * 5 = 45	**42.** 7 * 7 = 49
13. 6 * 9 = 54	**28.** 7 * 3 = 21	**43.** 9 * 11 = 99
14. 8 * 2 = 16	**29.** 3 * 11 = 33	**44.** 3 * 12 = 36
15. 2 * 10 = 20	**30.** 0 * 9 = 0	**45.** 4 * 4 = 16

Use with or after Lesson 1·3.

Practice Set 2 continued

SRB 208 222 249

46. ✏️ **Writing/Reasoning** Is 8 a factor of 70? Explain how you found your answer.

_____

_____

COMPUTATION PRACTICE **Solve.**

47. 250
       × 4
       1000

48. 525
       − 78
       447

49. 184
       − 67
       117

50. 21
       + 55
       76

51. 35
       + 24
       59

52. 39
       + 16
       55

53. 19
       × 30
       +4 78
       478

54. 531
       + 721
       1252

55. 1,800
       − 744
       1056

56. (60 + 30) * 5 = _____

57. 38 − (6 * 6) = _____

58. 48 + 22 + 27 = _____

59. $\left(\frac{63}{9}\right) * 7 =$ _____

Round to the nearest thousand.

60. 47,983 ___48,000___

61. 5,255 ___5,000___

62. 76,529 ___77,000___

63. 383,051 ___383,000___

Write the coordinates of the points shown on the coordinate grid.

64. A ___0,4___

65. B ___1,0___

66. C ___5,2___

67. D ___2,2___

68. E ___4,4___

69. F ___4,1___

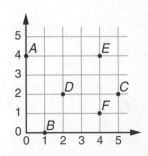

Practice Set 3

Use the numbers from 30 to 50 to answer Problems 1–4.

1. List the even numbers. _____

2. List the odd numbers. _____

3. List the numbers that have 5 as a factor. _____

4. List the numbers that have 4 as a factor. _____

FACTS PRACTICE **Solve.**

5. $5 + 4 =$ _____

6. $12 - 4 =$ _____

7. $10 + 7 =$ _____

8. $11 - 5 =$ _____

9. $9 + 6 =$ _____

10. $11 - 8 =$ _____

Write each of the following in dollars-and-cents notation.

11. 4 quarters, 3 dimes, 2 nickels, 4 pennies _____

12. 6 quarters, 6 dimes, 1 nickel, 2 pennies _____

---

13. **Writing/Reasoning** Kim is given a fact triangle with the numbers 48, 8, and 6. She must write the fact family for this triangle. Kim writes $48 \div 6 = 8$ as one of her facts. Is this answer correct? Explain your answer.

_____

_____

_____

Use with or after Lesson 1·4.

Practice Set 4

Use divisibility rules to help you answer these questions.
Write *yes* or *no*.

1. Is 300 divisible by 5? _____

2. Is 752 divisible by 2? _____

3. Is 5,225 divisible by 3? _____

4. Is 39,105 divisible by 9? _____

5. Is 18,373 divisible by 5? _____

6. Is 103,748 divisible by 10? _____

Write the numbers in order from least to greatest.

7. $\frac{7}{10}, \frac{5}{10}, \frac{9}{10}, \frac{3}{10}, \frac{6}{10}$ _____

8. $\frac{3}{5}, \frac{3}{7}, \frac{3}{4}, \frac{3}{2}, \frac{3}{8}$ _____

9. $\frac{1}{2}, \frac{3}{5}, \frac{1}{4}, \frac{7}{10}, \frac{11}{12}$ _____

Complete the name-collection boxes. Use as many different numbers and operations as you can.

Example

189
(60 * 3) + 9
378 / 2
(200 − 15) + 4

10.

245

11.

138

12.

79

13.

402

Practice Set 4 continued

Complete.

14. $10^3 =$ _1600_

15. $10^{\boxed{4}} = 10,000$

16. $10 * 10 * 10 * 10 * 10 =$ _100,000_

17. 10 to the eighth power = _100,000,000_

Rewrite the number models with parentheses to make them correct.

18. $7 * 9 - 4 = 35$ _____

19. $7 * 9 - 4 = 59$ _____

20. $32 - 16 - 7 = 9$ _____

21. $32 - 16 - 7 = 23$ _____

22. $589 = 6 * 25 + 75 - 11$ _____

Solve.

23. How many 7s are in 7,700? _____

 Solve.

24. 2,000
    * 43
   80,000

25. 150
   * 30
  4500

26. 2,500
    * 5
  12,500

27. 175
   * 20
  3500
  3500

28. 428
   * 8
  3424

29. 92
  * 15
  460
+ 920
1,380

30. 800
  * 45
36,000

31. 750
  * 18
  3900
+ 7500

32. 6)78

33. 30)4,500

34. $\frac{450}{9} = 50$

35. 7)378

36. A soup company offered to donate 40¢ for every soup-can label a school turned in. Forty-three students at Audubon School each brought in 25 soup-can labels. How many soup-can labels did all the students bring in? _____

37. How much will the soup company donate to Audubon School? _____

Practice Set 5

SRB
12 189

Write *prime* or *composite* for each number.

1. 18 _____   **2.** 17 _____   **3.** 39 _____

4. 43 _____   **5.** 50 _____   **6.** 23 _____

7. 42 _____   **8.** 77 _____   **9.** 37 _____

10. **Writing/Reasoning** How do you know if a number is prime?

_____

_____

_____

Find the area of each rectangle and write the number model.

Example

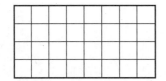

4 * 8 = 32 square units

Area = length (*l*) × width (*w*)

11.

12.

_____   _____

13.

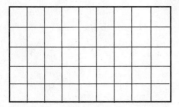

14.

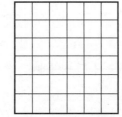

_____   _____

Use with or after Lesson 1·6.

19

Practice Set **5** *continued*

Write each decimal as a percent.

15. 0.40 _____

16. 0.25 _____

17. 0.12 _____

18. 0.85 _____

19. 1.67 _____

20. 0.05 _____

Complete the number lines.

21.

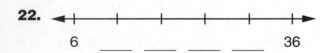

8 ____ ____ 32 ____ ____ ____ 64

22.

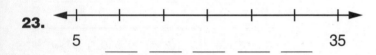

6 ____ ____ ____ ____ 36

23.

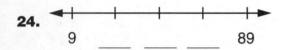

5 ____ ____ ____ ____ ____ 35

24.

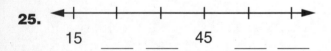

9 ____ ____ ____ 89

25.

15 ____ ____ 45 ____ ____

26.

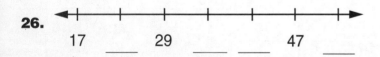

17 ____ 29 ____ ____ 47 ____

27.

−1 ____ ____ $\frac{1}{2}$ ____ $1\frac{1}{2}$

Use with or after Lesson 1•6.

Practice Set 6

Write the letter of the array that matches each square number.

1. 36 _____   **2.** 64 _____   **3.** 16 _____   **4.** 9 _____

A.

B.

C.

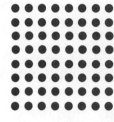

D.

List all the factors of each number. Tell whether each number is *prime* or *composite*.

5. 24 _____   **6.** 50 _____

7. 17 _____   **8.** 44 _____

Compare. Write < or >.

9. 33,085 _____ 13,058   **10.** 41,123 _____ 13,058

11. 110,362 _____ 101,317   **12.** 583,627 _____ 588,267

Write the digit in the thousands place.

13. 71,345 _____   **14.** 836,210 _____   **15.** 9,219 _____

16. 415,740 _____   **17.** 307,912 _____   **18.** 1,927,435 _____

Solve.

19.   444
    $\times$ 80

20.   35
    $\times$ 3

21.   749
   $-$ 482

22. $4\overline{)196}$

Practice Set 6 *continued*

Write each of the following in standard notation.

23. 4^2 _____

24. 12^2 _____

25. 27^2 _____

26. 25^2 _____

27. 40^2 _____

28. 62^2 _____

Six people share $111 equally.

29. How many $100 bills does each person get? _____

30. How many dollars are left to share? _____

31. How many $10 bills does each person get? _____

32. How many dollars are left to share? _____

33. How many $1 bills does each person get? _____

34. How many dollars are left over? _____

35. If the leftover money is shared equally,
how many cents does each person get? _____

36. Write a number model for the above problem. _____

COMPUTATION PRACTICE   **Solve.**

37. $210 - 180 =$ _____

38. $526 + 127 =$ _____

39. $80 + 36 =$ _____

40. $52 - 17 =$ _____

41. $97 - 8 =$ _____

42. $90 * 9 =$ _____

43. $587 - 236 =$ _____

44. $2,662 - 141 =$ _____

45. $370 * 8 =$ _____

46. $262 + 3,455 =$ _____

47. $120 * 50 =$ _____

48. $2,625 + 5,213 =$ _____

Use with or after Lesson 1•7.

Practice Set 7

Write the letter that matches the square root of each number.

1. 49 _____   **A.** 0.5

2. 0.25 _____   **B.** 13

3. 169 _____   **C.** 20

4. 400 _____   **D.** 7

Tell whether each number is a square number. Write *yes* or *no*.

5. 64 _____   **6.** 177 _____   **7.** 90 _____

8. 144 _____   **9.** 225 _____   **10.** 250 _____

Write the number sentences with parentheses. Then solve.

11. Add 70 to the difference of 365 and 36.

_____

12. Subtract the sum of 24 and 13 from 48.

_____

13. Add 7 to the difference of 37 and 15.

_____

14. Subtract the sum of 18 and 222 from 428.

_____

Write the following numbers in digits.

15. eighty million, three hundred twenty-one thousand, nine hundred eleven

_____

16. two billion, fifty-six thousand, five hundred

_____

17. six hundred fourteen billion, three hundred million

_____

Practice Set 8

SRB
6 12
128

Rewrite each product, using exponents.

1. 3 * 3 * 3 _____

2. 3 * 3 * 5 * 5 * 5 _____

3. 2 * 2 * 7 * 7 _____

4. 5 * 5 * 7 * 7 _____

5. 3 * 3 * 11 * 11 _____

6. 5 * 5 * 5 * 5 _____

Circle the prime factorization for each number.

7. 16   2 * 2 * 2 * 2   2 * 2 * 2 * 4

8. 9   2 * 3   3 * 3

9. 30   3 * 10   2 * 3 * 5

Suppose you spin a paper clip on the spinner shown.
Write *true* or *false* for each statement.

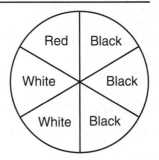

10. The paper clip is most likely to land on black. _____

11. The paper clip is least likely to land on red. _____

12. The paper clip is equally likely to land on black as on white. _____

COMPUTATION PRACTICE

Solve.

13.   20
       + 17

14.   31
       − 15

15.   38
       + 24

16.   320
       − 160

17.   560
       + 481

18.   745
       − 260

Practice Set 8 *continued*

 For each set of problems, do as many as you can in one minute. You can ask someone to time you.

Problem Set 1	Problem Set 2	Problem Set 3
19. $\frac{108}{9}$ = _____	**34.** $9 * 6$ = _____	**49.** $6 * 6$ = _____
20. $\frac{44}{4}$ = _____	**35.** $7 * 7$ = _____	**50.** $12 * 11$ = _____
21. $\frac{121}{11}$ = _____	**36.** $12 * 8$ = _____	**51.** $8 * 4$ = _____
22. $\frac{90}{9}$ = _____	**37.** $2 * 10$ = _____	**52.** $3 * 8$ = _____
23. $\frac{132}{12}$ = _____	**38.** $11 * 4$ = _____	**53.** $4 * 10$ = _____
24. $12 * 5$ = _____	**39.** $12 * 6$ = _____	**54.** $\frac{64}{8}$ = _____
25. $10 * 11$ = _____	**40.** $3 * 11$ = _____	**55.** $\frac{49}{7}$ = _____
26. $12 * 7$ = _____	**41.** $12 * 9$ = _____	**56.** $\frac{144}{12}$ = _____
27. $9 * 7$ = _____	**42.** $4 * 7$ = _____	**57.** $\frac{21}{3}$ = _____
28. $4 * 12$ = _____	**43.** $11 * 11$ = _____	**58.** $\frac{55}{5}$ = _____
29. $6 * 7$ = _____	**44.** $\frac{81}{9}$ = _____	**59.** $\frac{63}{9}$ = _____
30. $4 * 3$ = _____	**45.** $8 * 8$ = _____	**60.** $\frac{21}{7}$ = _____
31. $\frac{14}{2}$ = _____	**46.** $12 * 12$ = _____	**61.** $15 * 3$ = _____
32. $17 * 2$ = _____	**47.** $\frac{16}{4}$ = _____	**62.** $\frac{48}{8}$ = _____
33. $12 * 10$ = _____	**48.** $6 * 9$ = _____	**63.** $9 * 8$ = _____

Name _____ Date _____ Time _____

Practice Set 9

 Solve.

1. 322 + 921 = _____

2. 22 + 42 + 14 = _____

3. 540 + 191 = _____

4. 76.271 + 3.109 = _____

5. 6.152 + 8.019 = _____

6. 6.2 + 3.9 + 4.5 = _____

7. 16.5 + 97 = _____

8. 3.58 + 65.4 = _____

---

9. Use the clues to find the mystery number.

_____ _____ _____ . _____ _____ _____

- Add 43 and 23. Divide by 11 and write the result in the ones place.

- Triple the number in the ones place and divide by 2. Write the result in the tenths place.

- Multiply 8 ∗ 9. Subtract 68. Write the result in the thousandths place.

- Subtract the number in the tenths place from 57 and divide by 6. Write the result in the hundredths place.

- Divide 36 by the number in the thousandths place. Write the result in the tens place.

- Subtract the number in the ones place from the number in the tens place. Write the result in the hundreds place.

10. **Writing/Reasoning** Mai multiplied 5 × 7 to find the area of this rectangle. Kyle multiplied 5 × (3 + 4) to find the area. Were both students correct? Explain your answer.

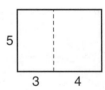

_____

_____

Use with or after Lesson 2•2.

Practice Set 9 continued

SRB
246

Eight people are going to share $682 equally.

11. How many $100 bills does each person get? _____

12. How many dollars are left to share? _____

13. How many $10 bills does each person get? _____

14. How many dollars are left to share? _____

15. How many $1 bills does each person get? _____

16. How many dollars are left over? _____

17. If the leftover money is shared equally,
how many cents does each person get? _____

18. Write a number model for the above problem. _____

Write the amounts.

19. (Q)(Q)(Q)(D)(D)(D)(N)(N)(P)(P)(P)(P) _____

20. [$1] [$1] [$1] (Q)(D)(D)(D)(D)(N)
(P)(P)(P) _____

21. [$5] [$5] [$5] [$5] [$1] (Q)(N)(N) _____

22. [$100] [$100] [$20] [$20] [$5]
[$1] [$1] (Q) _____

Fill in the missing numbers on the number lines.

23.
9   17   ___   33   ___   ___   57   ___

24.
0.2   ___   ___   1.0   1.2   ___   ___   1.8

Practice Set 10

SRB
15 19
182 183

 Solve.

1. 7.49 − 6.65 = _____ **2.** 4.8 − 1.2 = _____

3. 819 − 742 = _____ **4.** 346 − 122 = _____

5. 5.32 − 4.59 = _____ **6.** 9,007 − 3,568 = _____

7. 47.9 − 10.7 = _____ **8.** 5,300 − 1,792 = _____

Tell whether the measurement is *too small*, *too large*, or *OK*.

9. The book weighs 1 ton. _____

10. Ellen's sister is 5 yards tall. _____

11. The distance between San Francisco
and Washington, D.C. is 115 miles. _____

12. The chicken weighs about 4 pounds. _____

13. The nickel weighs about 5 grams. _____

14. A mug holds about 1 liter of water. _____

15. Mr. Brown's shoe is about 13 centimeters long. _____

Complete the missing factors.

16. 70 ∗ _____ = 210 **17.** _____ ∗ 4 = 360

18. _____ ∗ 8 = 640 **19.** 12 ∗ _____ = 960

20. 400 ∗ _____ = 3,600 **21.** _____ ∗ 50 = 350

22. 9 ∗ _____ = 810 **23.** _____ ∗ 6 = 6,600

28

Practice Set 11

Write a number sentence. Then find the solution.

1. The bakery sold 16 cherry pies and
   21 strawberry pies. How many pies were sold in all? _____

2. Sue paid for lunch with a $10 bill. The tuna
   sandwich cost $2.49 and the orange juice
   cost $1.65. How much change did she receive? _____

3. Evan spent $35.72 on school supplies. Tanya
   spent $23.18. How much more did Evan spend than Tanya? _____

Estimate the total cost.

4. 10 rulers that cost 79¢ each _____

5. 8 scissors that cost $1.14 each _____

6. 15 books that cost $2.35 each _____

7. 3 markers that cost $1.85 each _____

Use the figure at the right to answer the following questions.

8. How many sides does the polygon have?

   _____

9. What is the name of the polygon shown?

   _____

10. If each side is 2.4 cm, what is the perimeter?

   _____

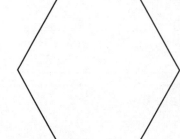

11. How many lines of symmetry does it have?

   _____

Practice Set 11 continued

12. Write the missing numbers. You may use a calculator.

Product	Exponential Notation	Standard Notation
5 * 5 * 5 * 5	5^4	625
8 * 8 * 8		
	10^5	
		144
13 * 13		
		400
	25^4	
30 * 30 * 30		

COMPUTATION PRACTICE **Find the square numbers.**

13. $9^2 =$ _____

14. $14^2 =$ _____

15. $20^2 =$ _____

16. $24^2 =$ _____

17. $32^2 =$ _____

18. $48^2 =$ _____

19. $65^2 =$ _____

20. $71^2 =$ _____

21. $100^2 =$ _____

22. $36^2 =$ _____

Complete.

23. $2\frac{1}{2}$ hours = _____ minutes

24. 12 hours = _____ day

25. 2,700 seconds = _____ minutes

26. $10\frac{1}{2}$ days = _____ weeks

27. 365 days = _____ year

28. 34 weeks = _____ days

29. 1 week = _____ hours

30. 336 hours = _____ days

31. 730 days = _____ years

32. 1 day = _____ minutes

Use with or after Lesson 2•4.

Practice Set 12

Suppose you spin a paper clip 200 times on the spinner below.
About how many times would you expect it to land on . . .

1. red? _____

2. blue? _____

3. yellow? _____

4. green? _____

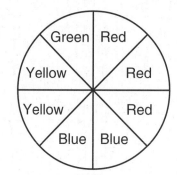

5. **Writing/Reasoning** Is the paper clip more likely to land
on green or on red? Explain your answer.

_____

_____

Write the digit in the hundredths place.

6. 2.15 = _____ **7.** 10.07 = _____ **8.** 3.142 = _____

9. 92.103 = _____ **10.** 7.13 = _____ **11.** 8.49 = _____

Write the fractional part for each picture.

Example

 $\frac{2}{12}$ or $\frac{1}{6}$

12.  _____

13.  _____

14.  _____

15.  _____

Use with or after Lesson 2•6.

Practice Set 12 continued

The following table shows when certain United States presidents were sworn into office. It also tells at what age each of these presidents were sworn in.

President	Date Sworn In	Age
Ford	August 9, 1974	61
Carter	January 20, 1977	52
Reagan	January 20, 1981	69
Bush	January 20, 1989	64
Clinton	January 20, 1993	46

16. What is the mean (average) age of the presidents at the time they were sworn in?

_____

17. Who was president for the shortest time?

_____

18. Presidents are elected for a term of 4 years. Which presidents served more than 1 term?

_____

19. Which of the presidents was the oldest when sworn in?

_____

20. How much older was Carter when he was sworn in as president than Clinton when he was sworn in?

_____

21. If President Ford was 61 when he was sworn in, how old was he on the same date in 1993?

_____

Use with or after Lesson 2•6.

Practice Set 13

Round each number to the nearest hundred.

1. 659 _____ **2.** 4,273 _____ **3.** 94.42 _____

4. 83,201 _____ **5.** 16,495 _____ **6.** 5,982 _____

Round each number to the nearest hundredth.

7. 4.023 _____ **8.** 15.617 _____ **9.** 0.179 _____

10. 6.005 _____ **11.** 732.424 _____ **12.** 95.189 _____

Make a magnitude estimate for the product. Is the solution in the
tenths, ones, tens, hundreds, thousands, **or** *ten-thousands***?**

13. 21 * 37 _____ **14.** 67 * 93 _____

15. 0.5 * 4.2 _____ **16.** 362 * 45 _____

17. 1.2 * 0.8 _____ **18.** 475 * 5.2 _____

 Solve.

19. 79
      + 356

20. 2,256
       − 36

21. 120
       − 30

22. 826
      + 182

23. 543
      + 768

24. 731
      + 610

25. 950
      − 635

26. 304
      − 170

27. 571
      + 39

Practice Set 14

 Solve.

1. 49 * 53 = _____

2. 19 * 247 = _____

3. 34 * 4.7 = _____

4. 3.2 * 9.7 = _____

5. 891 * 127 = _____

6. 27.5 * 16.8 = _____

7. 31 * 346 = _____

8. 16.2 * 97 = _____

9. 52.7 * 38 = _____

10. 6.1 * 25.4 = _____

11. 233 * 6.1 = _____

12. 37 * 5.8 = _____

Complete the "What's My Rule?" tables and the rule box.

13.

Rule
* 30

in	out
6	180
8	240
9	
11	
15	

14.

Rule

in	out
27	4
33	10
	14
24	
51	28

15.

Rule
+ 8.5

in	out
15	
23	
6	
41	
122	

16.

Rule
− 11

in	out
27	
33	
	14
24	
39	28

Use with or after Lesson 2•8.

Practice Set 14 *continued*

If 1 centimeter on a map represents 20 kilometers, then find the following:

17. 8 cm represents _____ km

18. 11 cm represents _____ km

19. 14 cm represents _____ km

20. 2.5 cm represents _____ km

21. 12 cm represents _____ km

Rewrite the number sentences with parentheses to make them correct.

22. $18 = 4 + 2 * 7$ _____

23. $33 - 14 - 5 = 24$ _____

24. $53 - 12 + 7 = 48$ _____

25. $8 * 9 + 4 * 12 = 120$ _____

26. $96 = 6 * 7 + 9$ _____

27. $4 * 3 + 8 * 10 = 440$ _____

28. $230 = 8 * 21 + 76 - 14$ _____

29. $8 * 10 - 3 = 56$ _____

Complete.

30. $10^2 =$ _____

31. $12 * 12 = 12^{\square}$

32. $9^{\square} = 81$

33. $5^{\square} = 25$

34. $2 * 2 = 2^{\square}$

35. $7^2 =$ _____

36. The square root of 121 = _____

37. The square root of 100 = _____

Practice Set 15

 Solve.

1. 61 * 24 = _____

2. 4.3 * 7 = _____

3. 92 * 1.37 = _____

4. 537 * 72 = _____

5. 18.3 * 6.5 = _____

6. 124 * 396 = _____

7. 4.15 * 2.7 = _____

8. 58 * 6.25 = _____

Tell whether each number is divisible by 3. Write *yes* or *no*.

9. 27 _____

10. 78 _____

11. 158 _____

12. 682 _____

Tell whether each number is divisible by 9. Write *yes* or *no*.

13. 36 _____

14. 93 _____

15. 117 _____

16. 487 _____

Solve.

17. _____ + 9 = 29

18. 300 + _____ = 500

19. 17 − _____ = 12

20. 100 − _____ = 75

21. 82 − _____ = 50

22. _____ + 92 = 108

23. _____ = 30 + 65

24. _____ − 60 = 55

25. 200 + _____ = 800

26. 170 − _____ = 90

Use with or after Lesson 2·9.

Practice Set ◁16▷

1. Use the clues to find the mystery number.

_____ _____, _____ _____ _____, _____ _____ _____, _____ _____ _____

- Find $\frac{1}{10}$ of 40. Double the result and write it in the thousands place.

- Double 4. Divide the result by 2. Write the answer in the ten-thousands place.

- Find 7 * 6. Reverse the digits in the result and divide by 8. Write the result in the millions place.

- Add 6 to the digit in the thousands place. Divide by 7 and write the result in the hundred-millions place.

- Write $\frac{12}{3}$ as a whole number in the ones place.

- Subtract the number in the ones place from the number in the ten-thousands place. Write the result in the hundreds place.

- Find $\frac{1}{3}$ of 18. Write the result in the hundred-thousands place.

- Find 20% of 35. Write the result in the ten-millions place.

- Add the number in the ten-millions place to the number in the hundred-millions place. Write the result in the ten-billions place.

- Find $\frac{1}{8}$ of 8 and write the result in the tens place.

- Find the sum of all the digits in the chart so far. Subtract 39 from the result and write the answer in the billions place.

2. Write the number in words.

_____

Practice Set 16 continued

Use words to write the following numbers.

3. 12,743,000 _____

4. 8,054,000,000,000 _____

5. 42,169,205,000,000 _____

6. 16,802,946 _____

7. Writing/Reasoning Explain how to square a number.

_____

_____

Complete.

8. $2^2 =$ _____

9. $4^{\square} = 16$

10. $5 * 5 = 5^{\square}$

11. 8 squared = _____

Write the number models with parentheses and solve.

12. Add 73 to the difference of 2,465 and 846. _____

13. Subtract the sum of 224 and 613 from 2,548. _____

14. Add 72 to the difference of 3,527 and 1,565. _____

15. Subtract the sum of 128 and 27 from 228. _____

Mr. Henderson reported these scores on a quiz:

   23, 18, 19, 23, 16, 12, 15, 12, 11, 16, 22, 19, 23

16. What is the maximum of the scores? _____

17. What is the minimum of the scores? _____

18. What is the range of the scores? _____

19. What is the mode of the scores? _____

20. What is the mean of the scores? _____

21. What is the median of the scores? _____

Use with or after Lesson 2•10.

Practice Set 17

Find the angle measures for the labeled angles. There are 360°
in a circle and 180° in a straight line. Do not use a protractor.

1. m∠G _____

2. m∠H _____

3. m∠F _____

4. m∠I _____

5. m∠J _____

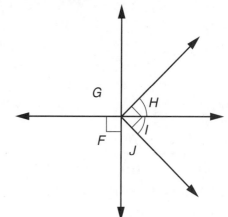

Write the letter of the best estimate for each problem.

6. 87 * 3.6 _____   **A.** 900

7. 2.87 * 3.6 _____   **B.** 2,000

8. 72 * 58 _____   **C.** 360

9. 879 * 1.17 _____   **D.** 12

10. 19.8 * 132.5 _____   **E.** 4,200

 Solve.

11.   524
        − 154

12.   426
        − 273

13.   684
        + 27

14. 2,647
        − 235

15.   700
        − 480

16.   526
        + 203

17.   609
        − 160

18.   222
        + 301

19.   458
        + 396

Practice Set **17** *continued*

SRB
17
138 139

Solve.

20. 464 + 2,079	**21.** 5,047 − 2,490	**22.** 5,844 − 2,398
23. 9,061 − 3,717	**24.** 4,096 + 2,833	**25.** 7,700 − 6,099

Find the angle measures for the labeled angles in the patterns below. Use the Geometry Template. Do not use a protractor.

26.

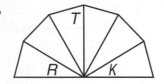

m∠K = _____
m∠R = _____
m∠T = _____

27.

m∠D = _____
m∠F = _____
m∠E = _____

28.

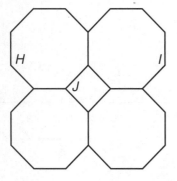

m∠H = _____
m∠I = _____
m∠J = _____

29.

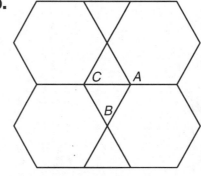

m∠A = _____
m∠B = _____
m∠C = _____

30. **Writing/Reasoning** Explain how you found the measure
of ∠I in Problem 28.

_____

_____

_____

40

Practice Set 18

Write the letter of the measurement that best describes each angle. Do not use a protractor.

1.  35°

A. 82°

2.  82°

B. 15°

3. 145°

C. 145°

4. 15°)

D. 35°

Write *true* or *false* for each number sentence.

5. 125 + 7 = 130 _false_

6. 23 * 6 > 100 _true_

7. 37.6 * 1.8 < 37.6 _false_

8. (5 * 2) + 18 = 41 _false_

9. 1.9 + 7.8 = 9.7 _True_

10. 1,600 ÷ 8 = 200 _true_

Round each number to the nearest hundredth.

11. 18.582 _18.58_

12. 5.826 _5.83_

13. 0.821 _0.82_

14. 634.624 _634.62_

15. 29.005 _29.01_

16. 23.205 _23.21_

Write the answer.

Express 25% as a fraction and as a decimal.

17. fraction _25/100_

18. decimal _0.25_

Practice Set **18** continued

SRB
4 57
222 223

Rewrite the number sentences with parentheses to make them correct.

19. $7 * 12 - 6 = 42$

_____

20. $7 * 12 - 6 = 78$

_____

21. $230 - 130 - 50 = 150$

_____

22. $21 = 3 * 2.4 + 4.6$

_____

23. $3 * 8.3 + 5 * 12 = 84.9$

_____

24. $300 = 5 * 70 - 50$

_____

25. $4 * 10 + 6 - 1 = 60$

_____

26. $378 = 12 * 30 + 18$

_____

Use digits to write the following numbers.

27. nineteen billion, six hundred million _____

28. seventy-six million, twenty thousand, six hundred fourteen _____

29. six trillion, four hundred fifty billion _____

Complete the number lines.

30.

$\frac{1}{6}$ ___ ___ ___ ___ $\frac{6}{6}$ or 1

31.

-15 ___ ___ ___ ___ ___ 15

32.

$-\frac{1}{4}$ ___ ___ $\frac{1}{2}$ ___ ___ ___

33.

-2 ___ -1 ___ 0 ___

34.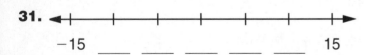

-10 ___ -2 ___ 6 ___

Use with or after Lesson 3•4.

Practice Set 19

Write *equilateral*, *isosceles*, or *scalene* to describe each triangle.

1.  equilateral

2. scalene

3.  isosceles

4. scalene

5. isosceles

6. equilateral

Use the clues to find the number.

7. Clue 1: I am a prime number less than 30.
   Clue 2: The sum of my digits is 5.

   _____ 14 _____

8. Clue 1: I am between 40 and 60.
   Clue 2: I am divisible by 5.
   Clue 3: The sum of my digits is 9.

   _____ 45 _____

9. Clue 1: I am greater than 500, but less than 1,000.
   Clue 2: My square root is a square number.

   _____ 900 _____

10. Clue 1: I am greater than 100,000, but less than 120,000.
    Clue 2: I am divisible by 3.
    Clue 3: The sum of my digits is 6.

    _____ 100,401 _____

Practice Set 20

Write the letters of all the names that match each figure.
Some figures have more than one name.

1. 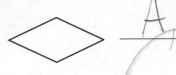 A, C   **A.** rhombus

2.  C   **B.** square

3.  B, C   **C.** quadrangle

4.  D   **D.** hexagon

Write the letter of the prime factorization for each number.

5. 16 E   **A.** $2 * 2 * 2 * 5$

6. 40 A   **B.** $2 * 2 * 3 * 3$

7. 18 D   **C.** $3 * 5 * 5$

8. 36 B   **D.** $2 * 3 * 3$

9. 75 C   **E.** $2 * 2 * 2 * 2$

 Solve.

10. $24 = 6 * $ __4__

11. $8.2 + 9.8 = $ __18.0__

12. $9 * $ __0.20__ $ = \$1.80$

13. __8500__ $/ 1,700 = 5$

14. $82.4 - 12.8 = $ __69.6__

15. $180 * $ __20__ $ = 3,600$

16. __9.80__ $/ 4.90 = 2$

17. $\$4.65 * 5 = $ __\$23.25__

18. $24.2 + 16 = $ __40.2__

19. $32.9 - 8 = $ __24.9__

Use with or after Lesson 3•7.

Practice Set 20 *continued*

Susan, Antonio, Chris, Jonathan, and Julie went out to dinner. They ordered 3 pizzas. All 3 pizzas were the same size.

20. Susan and Julie shared one pizza. Susan ate $\frac{3}{8}$ of the pizza. Julie ate $\frac{1}{2}$ of the pizza. Who ate more?

_____

21. How much of the pizza was left?

_____

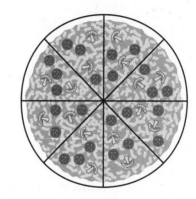

Antonio, Chris, and Jonathan shared the other two pizzas. Antonio ate $\frac{5}{8}$ of a pizza. Chris ate $\frac{3}{4}$ of a pizza. Jonathan ate $\frac{3}{6}$ of a pizza. Who ate more:

22. Chris or Antonio? _____

23. Chris or Jonathan? _____

24. Antonio or Jonathan? _____

25. Who ate the most? _____

26. How much was left? _____

27. How many slices were left from all 3 pizzas? _____

28. Susan, Antonio, Chris, Jonathan, and Julie want to share the remaining slices equally. How many sections should they divide the remaining slices into? _____

29. **Writing/Reasoning** Explain how you found the answer to Problem 28.

_____

_____

Practice Set 21

1. The hexagon can be divided into how many triangles? _____

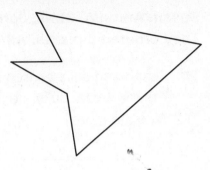

2. What is the sum of the angles in this hexagon? _____

3. The octagon can be divided Into how many triangles? _____

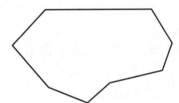

4. What is the sum of the angles of the octagon? _____

5. The dodecagon can be divided into how many triangles ? _____

6. What is the sum of the angles of the dodecagon? _____

7. **Writing/Reasoning** A 20-gon is a polygon with 20 sides. Explain how you could find the sum of the angles in a 20-gon.

_____

_____

_____

Use with or after Lesson 3•9.

Practice Set 21 continued

SRB
6 19 139

Write *acute, obtuse, right, straight,* or *reflex* to describe each angle.

8.   Right

9.   acute

10.  _____

11. _____

12. 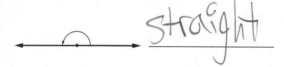 straight

13.  _____

COMPUTATION PRACTICE Solve.

14. $\frac{900}{90}$ = _____

15. 340 * 20 = _____

16. 1,200 = 120 * _____

17. 5 * 560 = _____

18. 40 * _____ = 6,000

19. $\frac{}{1,000}$ = 30

20. $\frac{2,800}{700}$ = _____

21. 16 * _____ = 560

22. $\frac{}{70}$ = 10

23. 9 * 30 = _____

24. $\frac{2,200}{}$ = 11

25. 8 * 60 = _____

Complete.

26. 10^8 = _____

27. $10^{\square}$ = 1,000,000,000

28. 10 * 10 * 10 * 10 * 10 * 10 * 10 = _____

29. 10 to the sixth power = _____

Practice Set 22

SRB
11 56
157

Use the figures below to answer Problems 1–4.

1. What fraction of the figures are triangles? _____ $\frac{2}{7}$

2. What fraction of the figures are polygons? _____ $\frac{5}{7}$

3. What fraction of the figures have at least 1 pair of parallel sides? _____ $\frac{3}{4}$

4. What fraction of the figures are not quadrangles? _____ $\frac{5}{7}$

Find the missing angle measure. A triangle has 180°.
Do not use a protractor.

5.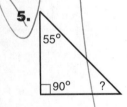

55°

90°   ?

$35°$

$\begin{array}{r} 90 \\ +55 \\ \hline 145 \end{array}$ $\begin{array}{r} 180 \\ -145 \\ \hline 35 \end{array}$

6. ?

125°   15°

175
140

$40°$

---

7. Sketch the reflection of the triangle below. Label the vertices
   of the triangle you draw.

B

A   C

Complete.

8. Underline the numbers that are divisible by 2. Circle the numbers
   that are divisible by 3. Cross out the numbers that are divisible by 5.

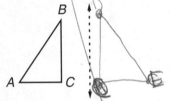

   340   845   9,303   1,001   653

   125   777   496   2,300   8,472

Practice Set 22 continued

FACTS PRACTICE For each set of problems, do as many problems as you can in one minute. You can ask someone to time you.

Problem Set 1	Problem Set 2	Problem Set 3
9. $7 * 1 = $ M	**24.** $9 + 1 = $ 10	**39.** $9 * 5 = $ 45
10. $\frac{21}{3} = $	**25.** $15 - 8 = $	**40.** $\frac{35}{7} = $
11. $\frac{28}{7} = $	**26.** $16 - 6 = $ 10	**41.** $8 + 2 = $ 10
12. $4 * 3 = $	**27.** $7 + 7 = $ 14	**42.** $16 - 9 = $
13. $\frac{18}{6} = $	**28.** $8 - 0 = $ 8	**43.** $8 * 5 = $ 40
14. $\frac{30}{3} = $	**29.** $6 + 7 = $ 13	**44.** $9 + 4 = $ 13
15. $5 * 0 = $ D	**30.** $5 + 4 = $ 9	**45.** $\frac{81}{9} = $
16. $\frac{24}{4} = $	**31.** $9 + 7 = $ 16	**46.** $16 - 7 = $
17. $\frac{30}{5} = $	**32.** $13 - 1 = $	**47.** $15 - 6 = $
18. $6 * 8 = $	**33.** $12 - 7 = $	**48.** $8 + 3 = $
19. $10 * 6 = $ 60	**34.** $3 + 5 = $ 8	**49.** $7 * 9 = $
20. $8 * 4 = $	**35.** $10 - 7 = $	**50.** $3 * 8 = $
21. $\frac{54}{9} = $	**36.** $2 + 0 = $ 2	**51.** $\frac{72}{8} = $
22. $6 * 6 = $ 36	**37.** $16 - 8 = $	**52.** $\frac{90}{9} = $
23. $\frac{56}{7} = $	**38.** $18 - 9 = $	**53.** $7 + 4 = $ 11

Practice Set 23

Use mental math to solve. Remember to break the number being divided into two or more friendly parts.

Example	**Friendly parts:**	**Divide each part.**
66 divided by 5	50 and 16	$50 \div 5 = 10$
		$16 \div 5 = 3$ with 1 left over
		66 divided by 5 equals **13 with 1 left over.**

1. 71 divided by 3 _____

2. 47 divided by 6 _____

3. 87 divided by 8 _____

4. 69 divided by 4 _____

5. 95 divided by 7 _____

6. 86 divided by 6 _____

Tell how many. Then write a number model.

7. How many dots are in this array? _____

. . . .
. . . .
. . . .
. . . .   **Number model:** _____

8. How many dots are in this array? _____

. . . .
. . . .
. . . .
. . . .
. . . .   **Number model:** _____

Complete.

9. $10^4 =$ _____

10. $5^{\square} = 25$

11. $10,000,000 = 10^{\square}$

12. $3^{\square} = 27$

13. $64 = 4^{\square}$

14. _____ $^4 = 81$

Use with or after Lesson 4•1.

Practice Set **23** *continued*

Write the value of the digit 8 in the numerals below.

15. 589,000 _____

16. 87,402,000,000 _____

17. 312,719,538 _____

18. 482,391,092 _____

19. 328,946,326 _____

20. Name the polygon shown below. _____

21. If each side were 6.9 centimeters, what would the perimeter be? _____

Complete the number lines.

22.

4.5 ___ ___ ___ ___ ___ 5.2

23.

−8 −6 ___ ___ ___ 2

24.

7 ___ ___ ___ ___ ___ 49

25.

3 ___ ___ ___ 123

Practice Set 24

Divide.

1. $\frac{518}{5} =$ _____

2. $183 \div 6 =$ _____

3. $464 \div 4 =$ _____

4. $\frac{630}{8} =$ _____

5. $\frac{967}{9} =$ _____

6. $\frac{1,344}{12} =$ _____

7. $6,568 \div 8 =$ _____

8. $3,068 \div 23 =$ _____

9. Ellen had 293 buttons. She places 6 buttons in each bag.
   How many bags of buttons can she make?

   _____

Tell whether each number is *even* or *odd*. Then list all of the factors.

10. 49 _____

11. 58 _____

12. 62 _____

13. 81 _____

14. 76 _____

15. 95 _____

Write the amounts.

16. $1 | $1 | $1 | Q D D D N P P _____

17. $5 | $1 | $1 | Q Q Q N N _____

18. $100 | $20 | $20 | $5 | $1 | $1 | $1 _____

Solve. If 1 in. represents 100 mi on a map, then

19. 2 in. represents _____ mi.

20. 3 in. represents _____ mi.

21. 10 in. represents _____ mi.

22. $\frac{3}{4}$ in. represents _____ mi.

Use with or after Lesson 4·2.

Practice Set 25

Use the map and map scale to answer the questions.

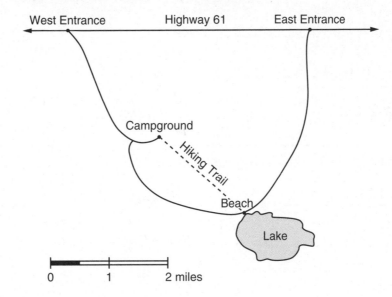

1. How far is it from the east entrance to the lake? _____

2. Which is closer to the beach, the east or west entrance?

   _____

3. Which entrance is closer to the campground? By how many miles?

   _____

4. Sue and Jason want to go from the campground to the beach.
   If Sue rides her bike on the road at 5 miles per hour, and Jason
   walks on the trail at 3 miles per hour, who will reach the beach first?

   _____

5. If there was a trail around the lake, estimate how long
   this trail would be.

   _____

6. **Writing/Reasoning** If you wanted to hike about 4 miles,
   describe a route you might take.

   _____

   _____

Practice Set 25 *continued*

Write the numbers in order from *least* to *greatest*.

7. 1.79, 0.12, 5.1, 0.4, 4.03 _____

8. 9.8, 0.98, 8.09, 8.9, 0.89 _____

9. 0.2, 2.2, 0.12, 1.2, 0.21 _____

Estimate the answer to each multiplication problem.

10. 185 * 22 = _____   **11.** 92 * 41 = _____

12. 781 * 68 = _____   **13.** 209 * 71 = _____

14. 314 * 18 = _____   **15.** 903 * 47 = _____

Solve.

16. 900 * 800 = p _____   **17.** 5,000 * d = 300,000 _____

18. 5,400 = x * 90 _____   **19.** 42,000 = 700 * s _____

20. 3 * 1,500 = n _____   **21.** $\frac{64,000}{8,000} = g$ _____

Solve.

22. A mole can dig a tunnel 300 feet long in one night. How many yards can a mole dig in three weeks? (Reminder: 3 ft = 1 yd)

23. A bottle-nosed dolphin can dive to a depth of 3,000 feet in 2 minutes. About how many yards per second is that?

24. When it snows, Shawn charges $4 for every sidewalk he shovels, and $5 for every driveway he shovels. If he shovels 8 sidewalks and 3 driveways, how much does he earn?

Practice Set 26

Make a magnitude estimate of the quotient. Is the solution in the *tenths, ones, tens,* or *hundreds?* Then find the exact answer.

1. $18.9 \div 7$ _____

2. $297 \div 5$ _____

3. $61.6 / 4$ _____

4. $25.2 / 6$ _____

5. $\$40.43 \div 3$ _____

6. $786 / 6$ _____

 Solve.

7. $623 + 812 =$ _____

8. $170 - 68 =$ _____

9. $495 - 381 =$ _____

10. $2,791 + 342 =$ _____

11. $3,465 + 1,273 =$ _____

12. $7,514 - 2,356 =$ _____

Write a number sentence. Then find the solution.

13. There are 17 cards in each box. There are 9 boxes on the shelf. How many cards are on the shelf? _____

14. The library is open 6 days a week. Each day, an average of 430 books are checked out. What is the average total number of books that are checked out in a week? _____

15. Karen bought a jacket for $42.59 and a pair of slacks for $23.65. How much did she spend in all? _____

16. **Writing/Reasoning** Explain how you found the answer to Problem 14.

_____

_____

_____

Practice Set 27

 Writing/Reasoning Write a number sentence, and then solve. Explain what the remainder represents.

1. 437 students need calculators. Calculators come in boxes of 12. How many boxes of calculators need to be ordered so that each student will have a calculator?

_____

_____

2. Ms. Woods has 27 feet of fabric. She needs to make 4 identical costumes. How much fabric does she have for each costume?

_____

_____

3. Oscar is making fruit baskets. Each fruit basket must have 15 pieces of fruit. How many baskets can he make with 123 pieces of fruit?

_____

_____

Rewrite the number sentences with parentheses to make them correct.

4. $42 * 8 - 5 = 126$

_____

5. $9.5 = 6.3 + \frac{6.4}{2}$

_____

6. $260 - 240 - 6 = 26$

_____

7. $10 - 12 + 9 = -11$

_____

8. $8 * 6 + 9 * 5 = 600$

_____

9. $170 = 20 * 4 + 90$

_____

10. $7 * 4 + 10 * 13 = 158$

_____

11. $115.8 = 11.6 * 3 + 9 * 9$

_____

Use with or after Lesson 4•5.

Practice Set 27 *continued*

Measure each angle to the nearest degree.

12.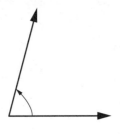

_____

13.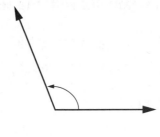

_____

Use digits to write the following numbers.

14. one hundred sixty-two and nine hundred seventy-four thousandths

_____

15. sixteen and one hundred forty-seven thousandths

_____

16. one thousand and forty-two thousandths

_____

17. sixteen and four hundred seven thousandths

_____

Use words to write the following numbers.

18. 171.603 _____

19. 34.087 _____

20. 1.042 _____

21. 71.627 _____

Practice Set 28

For each number story, write a number sentence and solve the
problem. Circle what you did with the remainder and explain why.

1. Eight people went to lunch. The bill came to $98.00. If they split the
    bill equally among them, how much did each person pay?

    Number sentence: _____

    Answer: _____

    Circle what you did about the remainder.

    Ignored it   Reported it as a fraction or a decimal   Rounded the answer up

    Why? _____

2. Rose collected 74 eggs from the chickens on her farm. She wants to
    place a dozen eggs in each carton. How many cartons will she need?

    Number sentence: _____

    Answer: _____

    Circle what you did about the remainder.

    Ignored it   Reported it as a fraction or a decimal   Rounded the answer up

    Why? _____

3. Andy has $57.00. He wants to buy 6 CDs for $9.00 each, including
    tax. Does he have enough money?

    Number sentence: _____

    Answer: _____

    Circle what you did about the remainder.

    Ignored it   Reported it as a fraction or a decimal   Rounded the answer up

    Why? _____

Practice Set 29

1. How many pieces of fruit are there altogether? _____

2. What fraction of the fruit are apples? _____

3. What fraction of the fruit are pears? _____

4. What fraction of the fruit are bananas? _____

5. What fraction of the fruit are oranges? _____

Write *true* or *false* for each number sentence.

6. $15 \div 3 < 16 \div 4$ _____

7. $2.93 - 1.05 = 2.43$ _____

8. $96 - (2 * 3) = 40 + 50$ _____

9. $375 \div 25 = 15$ _____

10. $17 + 23 = \frac{100}{2}$ _____

11. $0.5 * 6 = 6$ _____

Measure each line segment to nearest centimeter.

12. ├────────────────┤ _____ cm

13. ├──────────────────┤ _____ cm

14. ├──────────┤ _____ cm

Practice Set 29 continued

COMPUTATION PRACTICE **Solve.**

15. 32.76
      + 13.98

16. 3.187
      + 1.290

17. 547.2
      − 371.9

18. 30)‾180‾

19. 6)‾4.2‾

20. 9)‾103.5‾

21. 8,323
      + 1,475

22. 5,335
      + 3,182

23. 82,416
      + 15,249

24. A parking lot has 4 rows. Each row has spaces for 11 cars. How many cars can be parked in this lot? _____

25. Write a number model for the above problem.

_____

Fill in the missing numbers on the number lines.

26.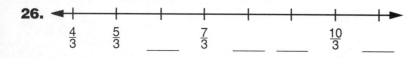

$\frac{4}{3}$ $\frac{5}{3}$ ___ $\frac{7}{3}$ ___   ___ $\frac{10}{3}$ ___

27.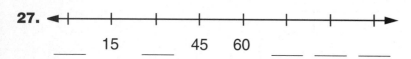

___   15   ___   45   60   ___   ___   ___

28.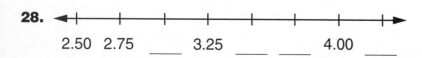

2.50   2.75   ___   3.25   ___   ___   4.00   ___

29.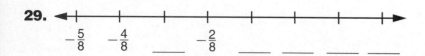

$-\frac{5}{8}$ $-\frac{4}{8}$ ___ $-\frac{2}{8}$ ___   ___   ___

30.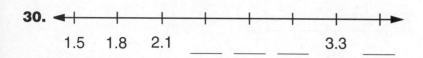

1.5   1.8   2.1   ___   ___   ___   3.3   ___

Use with or after Lesson 5·1.

Practice Set 30

Each whole circle below is $\frac{8}{8}$ or 1. Write the mixed number for each diagram.

1.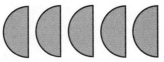

_____

2.

_____

3.

_____

4.

_____

5.

_____

6.

_____

Find the total cost of each of the following.

7. 18 pencils that cost 14¢ each _____

8. 7 scissors that cost $0.68 each _____

9. 4 books that cost $3.58 each _____

10. 28 rulers that cost $1.23 each _____

11. 2 pairs of shoes that cost $15.59 each _____

12. 8 pens that cost $0.98 each _____

13. 6 cans of juice that cost $4.24 each _____

14. 5 oranges that cost $0.20 each _____

15. 11 hats that cost $19.99 each _____

Practice Set 30 continued

Find the amounts.

16. Q Q Q Q Q D D N N P P P   _____

17. $1 $1 $1 Q D D D D P P   _____

18. $5 $5 $5 $5 $5 $1

Q N N   _____

19. $100 $100 $20 $20 $5 $1 $1   _____

Circle the correct answer.

20. What is the prime factorization for 42?

2 * 3 * 7   3 * 7 * 7

2 * 2 * 7   3 * 7

21. What is the prime factorization for 80?

2 * 4 * 5   2 * 2 * 2 * 2 * 5

2 * 2 * 2 * 3 * 5   5 * 5 * 2 * 2 * 2

22. What is the prime factorization for 34?

2 * 3 * 5   2 * 2 * 2 * 3

2 * 17   2 * 2 * 3 * 3

23. What is the prime factorization for 48?

2 * 5 * 5   2 * 2 * 2 * 2 * 3

2 * 23   2 * 2 * 2 * 3 * 3

Practice Set 31

Write two equivalent fractions for each fraction given.

1. $\frac{2}{3}$ _____

2. $\frac{5}{6}$ _____

3. $\frac{9}{12}$ _____

4. $\frac{4}{16}$ _____

5. $\frac{3}{10}$ _____

6. $\frac{1}{5}$ _____

Write the fractions in order from least to greatest.

7. $\frac{7}{12}$, $\frac{1}{12}$, $\frac{3}{12}$, $\frac{8}{12}$, $\frac{11}{12}$ _____

8. $\frac{3}{6}$, $\frac{3}{5}$, $\frac{3}{4}$, $\frac{3}{9}$, $\frac{3}{12}$ _____

9. $\frac{1}{4}$, $\frac{1}{8}$, $\frac{3}{16}$, $\frac{5}{8}$, $\frac{7}{16}$ _____

Write the number sentences with parentheses. Then solve.

10. Add 36 to the difference of 229 and 74. _____

11. Subtract the sum of 23 and 56 from 312. _____

12. Add 18 to the difference of 260 and 176. _____

13. Subtract the sum of 76 and 41 from 189. _____

Complete the number lines.

14.

$\frac{2}{9}$ ___ ___ ___ ___ $\frac{7}{9}$

15.

-42 ___ ___ ___ ___ ___ -30

16. If $\frac{1}{2}$ in. represents 10 mi on a map, then 10 in. represents _____ mi.

17. If $\frac{1}{2}$ in. represents 25 mi on a map, then 5 in. represents _____ mi.

Practice Set 32

SRB
59 60

Draw a horizontal line to divide each part of the fraction stick into 2 equal parts. Then fill in the missing numbers.

1. $\dfrac{3}{8} = \dfrac{6}{16}$

2. $\dfrac{3}{5} = \dfrac{6}{10}$

Draw horizontal lines to divide each part of the fraction stick into 3 equal parts. Then fill in the missing numbers.

3. $\dfrac{4}{6} = \dfrac{12}{18}$

4. $\dfrac{1}{5} = \dfrac{3}{15}$

Compare the fractions. Write = if they are equivalent. Write ≠ if they are not.

5. $\dfrac{2}{3}$ = $\dfrac{6}{9}$ **6.** $\dfrac{9}{16}$ ≠ $\dfrac{3}{4}$ **7.** $\dfrac{7}{10}$ ≠ $\dfrac{75}{100}$ **8.** $\dfrac{2}{7}$ = $\dfrac{6}{21}$

Fill in the boxes to complete the equivalent fractions.

9. $\dfrac{1}{3} = \dfrac{5}{15}$ **10.** $\dfrac{4}{16} = \dfrac{1}{4}$ **11.** $\dfrac{45}{60} = \dfrac{15}{20}$ **12.** $\dfrac{5}{9} = \dfrac{25}{45}$

Use the division rule to find equivalent fractions.

13. $\dfrac{20}{24} = \dfrac{5}{6}$ **14.** $\dfrac{9}{12} = \dfrac{3}{4}$ **15.** $\dfrac{3}{30} = \dfrac{1}{10}$ **16.** $\dfrac{36}{54} = \dfrac{2}{3}$

17. $\dfrac{6}{8} = \dfrac{3}{4}$ **18.** $\dfrac{6}{51} = \dfrac{2}{17}$ **19.** $\dfrac{51}{99} = \dfrac{17}{33}$ **20.** $\dfrac{333}{333} = \dfrac{1}{1}$

21. $\dfrac{21}{56} = \dfrac{3}{8}$ **22.** $\dfrac{16}{52} = \dfrac{4}{13}$ **23.** $\dfrac{18}{45} = \dfrac{2}{5}$ **24.** $\dfrac{42}{45} = \dfrac{14}{15}$

Use with or after Lesson 5•4.

Practice Set 33

Rename the shaded part of the square as a fraction and a decimal.

1.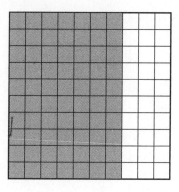

_____

2.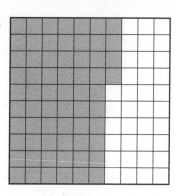

_____

3.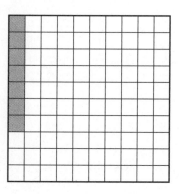

_____

4.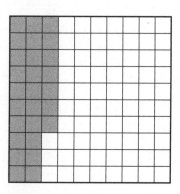

_____

5.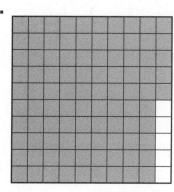

_____

6.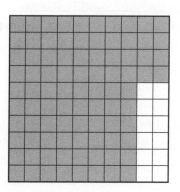

_____

Practice Set 33 continued

Use the map and map scale to answer the questions below.

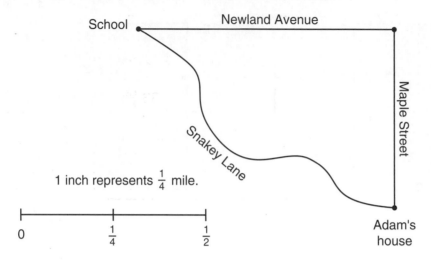

1 inch represents $\frac{1}{4}$ mile.

7. About how far would Adam have walked from school once he reached the corner of Newland and Maple?

_____

8. About how far would Adam have walked from school if he had taken Snakey Lane home?

_____

9. If Adam walked home from school along Snakey Lane at a rate of 2.5 miles per hour, and his school bus normally takes 25 minutes to drive him home, which mode of transportation would be faster? By how many minutes?

_____

10. If Adam's mother jogged from the house to the school along Snakey Lane and came back along Newland Avenue and Maple Street, about how far would she have jogged?

_____

11. If Adam lived on Maple Street, halfway between Snakey Lane and Newland Avenue, which route to school would be shorter?

_____

Use with or after Lesson 5·5.

Practice Set 34

SRB
45 89 90
208

1. Write the missing numbers for the table.

Fraction	Decimal	Percent
		50%
		$33\frac{1}{3}$%
	0.6	
	0.32	

Round to the nearest hundredth.

2. 14.096 _____

3. 3.746 _____

4. 2.149 _____

5. 23.692 _____

6. 4.385 _____

7. 5.001 _____

Write the coordinates of the points on the coordinate grid.

8. A _____

9. B _____

10. C _____

11. D _____

12. E _____

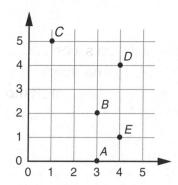

13. Using the same coordinate grid, write the coordinates of a point that would lie on the same line as points *B* and *D*.

_____

14. Using the same coordinate grid, write the coordinates of a point that would lie on the same line as points *A* and *E*.

_____

Practice Set 34 continued

COMPUTATION PRACTICE Solve.

15. 63
* 36

16. 95
* 26

17. 67
* 25

18. 17
* 57

19. 34,344
− 3,640

20. 80
* 17

21. 387
+ 3,643
  4,030

22. 40
* 70

23. 4,751
− 800
  3,951

24. 760
* 32

25. 52
* 19

26. 75
* 23

Write the amounts.

27. 3 $1, 1 Q, 3 D, 1 P _____ $3.56 _____

28. 3 $5, 2 $1, 1 Q, 3 N _____ $17.40 _____

29. 2 $100, 2 $20, 1 $5, 3 $1 _____ $248.00 _____

30. 7 Q, 2 D, 3 N, 3 P _____ $2.16 _____

31. 10 Q, 12 D, 18 N, 16 P _____ $4.70 _____

32. 12 $1, 16 Q, 44 N _____ $18.20 _____

Write the next three numbers in the pattern.

33. 18; 1,800; 180,000; _18,000,000_ ; _1,800,000,000_ , _180,000,000,000_

34. 53, 67, 81, _95_ , _169_ , _123_

35. 19, 8, −3, _−14_ , _−25_ , _−36_

36. 0.75, 15, 300, _6000_ , _120,000_ , _2,400,000_

Practice Set 35

Write the decimal as a fraction or a mixed number.

1. 0.7 $\frac{7}{10}$

2. 0.3 $\frac{3}{10}$

3. 1.5 $1\frac{5}{10}$

4. 4.2 $4\frac{2}{10}$ $4\frac{1}{5}$

5. 2.9 $2\frac{9}{10}$

6. 2.7 $2\frac{7}{10}$

7. 0.51 $\frac{51}{100}$

8. 8.13 $8\frac{13}{100}$

9. 12.07 $12\frac{7}{100}$

Write *yes* if the fractions are equivalent. Write *no* if they are not.

10. $\frac{10}{12}, \frac{5}{6}$ yes

11. $\frac{5}{10}, \frac{2}{3}$ no

12. $\frac{3}{21}, \frac{1}{7}$ yes

13. $\frac{9}{24}, \frac{3}{8}$ yes

14. $\frac{2}{6}, \frac{3}{5}$ No

15. $\frac{2}{5}, \frac{10}{15}$ No

16. $\frac{14}{20}, \frac{2}{7}$ No

17. $\frac{3}{12}, \frac{12}{48}$ yes

Find the perimeter of each shape.

18.

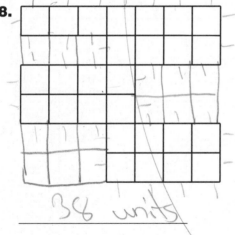

38 units

19.

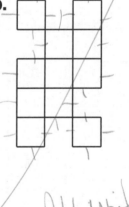

24 units

20. **Writing/Reasoning** Describe two ways to find the area of the shape in Problem 18.

Just count the squares inside the shape.

Practice Set 36

Find the landmarks for the data of the number of pages read
by each student as shown on the graph.

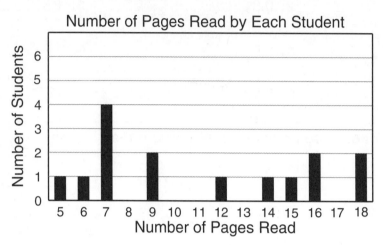

Number of Pages Read by Each Student

1. minimum _____

2. maximum _____

3. range _____

4. mode _____

5. mean _____

6. median _____

Complete the "What's My Rule?" tables.

7.

Rule		in	out
out = in / 25		300	
		475	
		825	
			17
		150	

8.

Rule		in	out
out = in / 9		270	
		81	
			14
		117	
			21

Solve.

9. $636 - x = 85$ _____

10. $15.9 + 38.5 = t$ _____

11. $152 + 652 = p$ _____

12. $847 - 264 = d$ _____

13. $91 - s = 44$ _____

14. $64 + d = 138$ _____

Use with or after Lesson 5·9.

Practice Set 36 continued

Use the information below to answer the questions and complete the circle graph.

The school cafeteria surveyed 60 students to find what most students prefer to drink with their lunch. Half of the students surveyed said they preferred milk. 15 students said water was their favorite drink, and 9 students reported that orange juice was their favorite. The rest of the students surveyed said they preferred apple juice.

15. How many students preferred milk? _____

16. What percent of the students preferred milk? _____

17. What percent of the students preferred water? _____

18. What percent of the students preferred orange juice? _____

19. How many students preferred apple juice? _____

20. What percent of the students preferred apple juice? _____

21. Label each section of the graph with the type of drink and the percent of students who prefer that drink. Title your graph.

_____
(Title)

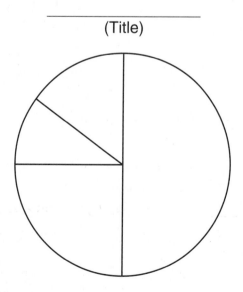

Practice Set 37

Use the circle graph to answer the questions below.

Favorite Sports

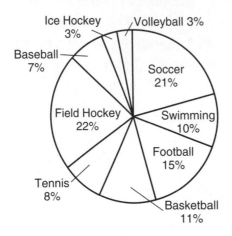

1. Which is the favorite sport of students in this survey? _____

2. Which is more popular, tennis or baseball? _____

3. About how many times more popular is field hockey than swimming? _____

4. How many times more popular is soccer than ice hockey? _____

5. What percent of the students like sports that use a spherical ball? _____

6. What percent of the students like sports in which you hit a ball? _____

7. What percent of the students like sports in which you throw a ball? _____

8. About how many times more popular is football than tennis? _____

9. **Writing/Reasoning** Write your own question, using the data from the circle graph. Then write the answer.

_____

_____

Use with or after Lesson 5·10.

Practice Set 37 *continued*

Keiko made 12 of 16 shots in the basketball game.

10. What fraction of the shots did she make? _____

11. What percent of the shots did she make? _____

12. At this rate, how many shots would she make if she took 20 shots? _____

Mr. Ryan set a goal of running a total of 100 miles each month. He filled in the squares on the grid as shown to keep track of the number of miles he ran. Each square represents 1 mile.

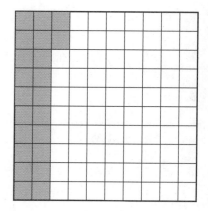

13. How many miles has Mr. Ryan run so far this month?

_____

14. What fraction of 100 miles did he run so far?

_____

15. What percent of his goal has he reached?

_____

16. If it took him 10 days to run this many miles, do you think he will reach his goal?

_____

17. About how many miles should he run each day in an average month in order to reach his goal?

_____

18. About how many miles will he need to run the rest of the days in this month in order to reach his goal?

_____

Practice Set 38

The graph below shows the number of books students read, outside of school, in one month. Find the landmarks for the data.

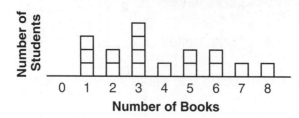

1. minimum _____

2. maximum _____

3. median _____

4. mean _____

5. mode _____

6. range _____

 Solve.

7. 2,940
   + 185

8.   59
   * 28

9. 7,474
   + 2,852

10. 2,058
   − 1,744

11. 10,769
   + 2,375

12.   63
   * 36

13. 4.928
   − 1.878

14. 2.564
   − 1.9

15. 9.375
   + 2.058

Complete.

16. 6 ft = _____ in.

17. 2 yd = _____ ft

18. 14 ft 3 in. = _____ in.

19. 7 yd 4 ft = _____ ft

20. 76 in. = _____ ft _____ in.

21. 28 ft = _____ yd _____ ft

22. 3,160 in. = _____ yd _____ ft _____ in.

Use with or after Lesson 6•1.

Practice Set 39

Measure each line segment to the nearest millimeter.

1. |———| _____ mm

2. |————————| _____ mm

3. |————————| _____ mm

4. |——| _____ mm

5. Use the clues to find the mystery number.

_____ . _____ _____ _____

- Multiply 6 ✳ 40. Subtract 233 and write the result in the ones place.

- Find 75% of 4. Write the result in the hundredths place.

- Find $\frac{1}{20}$ of 100. Write the result in the thousandths place.

- Add 5 to the number in the hundredths place. Write the result in the tenths place.

Circle the numbers that are divisible by 2 *and* 3.

6. 3,411   **7.** 3,846   **8.** 8,036   **9.** 552

10. 9,992   **11.** 144   **12.** 603   **13.** 7,212

Complete the circle graph. Give the graph a title.

14. Ms. Li took a survey to see what kinds of pets her fifth-grade students had. She found that 50% of her students had dogs, 30% had cats, 15% had no pets, and 5% had hamsters. Use the results of Ms. Li's survey to draw and label the circle graph.

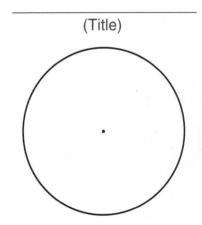

_____
(Title)

Practice Set 40

Use the stem-and-leaf plot to answer Problems 1–4.

1. How many scores are reported
on the stem-and-leaf plot?

_____

2. What is the maximum?

_____

3. What is the minimum?

_____

4. What is the median?

_____

Science Test Scores

Stems (10s)	Leaves (1s)
2	7 7 8
3	1 3 3 5 6 8 8 9
4	0 0 1 2 3 3 3 3 4 5 7
5	0 0

Tell whether each triangle is *equilateral, isosceles,* or *scalene.*

5.

 _____

6.

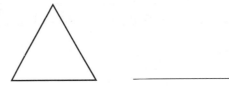 _____

7.

 _____

8.

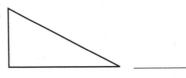 _____

Compare. Write < or >.

9. 265,168 _____ 29,518

10. 51,462 _____ 54,169

11. 1,645,283 _____ 1,644,823

12. 22,469,743 _____ 22,567,843

 Solve.

13.   8,003
− 2,694

14.   3,094
− 1,507

15.   6,999
− 4,025

16.   5,219
+ 5,897

17.   6,420
+ 4,824

18.   3,715
+ 7,091

Use with or after Lesson 6·3.

Practice Set **40** *continued*

 **Solve.**

19. 65 * 65 = _____

20. 345 / 15 = _____

21. 835 * 3 = _____

22. 1500 / 25 = _____

23. 23 * 25 = _____

24. 169 / 13 = _____

25. 39 * 2 = _____

26. 650 / 15 = _____

> 1 gallon = 4 quarts = 16 cups

27. A good milking cow will give up to 6,000 quarts of milk in a year. How many gallons is that?

_____

28. About how many gallons is that per day?

_____

29. If a family uses 2 gallons of milk per week, how many cups of milk does the family consume in a year?

_____

30. About how many quarts will a family use in 6 months?

_____

Complete the name-collection box for each number below.
Use as many different numbers and operations as you can.

31.
419.641

32.
3.805

33.
34.152

Practice Set 41

SRB
4 117

Examine the mystery plots below. Write the units.

Plot #1 Unit: _Ages of people. Retiring from a business_

```
                          x
                  x   x   x
                  x   x   x   x       x           x
      x   x       x   x   x   x   x   x   x       x
```
| 52 | 53 | 54 | 55 | 56 | 57 | 58 | 59 | 60 | 61 | 62 | 63 | 64 | 65 | 66 |

Plot #2 Unit: _Ages Of Fifth Graders' Mothers_

```
                          x
              x           x
              x   x   x   x       x
      x       x x     x x x   x x x   x   x   x   x   x   x   x
```
| 26 | 28 | 30 | 32 | 34 | 36 | 38 | 40 | 42 | 44 | 46 | 48 | 50 | 52 | 54 |

1. Which graph could describe the ages of fifth graders' mothers?

   Plot #2

2. Which graph could describe the ages of people retiring from a business?

   Plot #1

In the numeral 7,128,490,563 the 8 stands for 8,000,000.

3. What does the 7 stand for? _7,000,000,000_

4. What does the 1 stand for? _100,000,000_

5. What does the 0 stand for? _0,563_

6. What does the 9 stand for? _90,000_

7. What does the 2 stand for? _20,000,000_

8. What does the 4 stand for? _400,000_

Use with or after Lesson 6•4.

Practice Set 41 continued

Write the digit in the hundredths place.

9. 5.392 _____

10. 3.731 _____

11. 0.027 _____

12. 1.856 _____

13. 8.374 _____

14. 702.6152 _____

Write the next three numbers in the pattern.

15. 60, 180, 300, _____, _____, _____

16. 9, 15, 21, _____, _____, _____

17. $\frac{6}{4}, \frac{5}{4}, \frac{4}{4}$, _____, _____, _____

Write the amounts.

18. | $1 | $1 | Q | D | N | N | N | N | P | _____

19. | $5 | $5 | $5 | $5 | $5 | $5 | $1 |
| $1 | $1 | $1 | Q | Q | N | _____

20. | $100 | $20 | $20 | $20 | $5 | $1 |
Q Q Q   _____

21. Measure and label the sides of the shape to the nearest $\frac{1}{10}$ cm.

22. Write a number model to find the perimeter of the shape.

Number model: _____

_____ cm

_____ cm

23. What is the perimeter? _____

24. Write a number model to find the area of the shape.

Number model: _____

25. What is the area? _____

Practice Set 42

SRB
47–50

For each number story, tell the unit and the whole.
Answer each question, using a fraction and a percent.

1. In the pet store, there are 9 terriers, 6 beagles, and
5 poodles for sale. What fraction of the dogs are poodles?

Unit: _____   Whole: _____

Fraction: _____   Percent: _____

Total		
?		
Part	Part	Part
9	6	5

2. The Bags for You store tracks its sales for one week.
The store sold 15 bags on Monday. On Tuesday, 50 bags
were sold. The store sold 35 bags on Wednesday. What
fraction of the bags were sold on Monday?

Unit: _____   Whole: _____

Fraction: _____   Percent: _____

Total		
?		
Part	Part	Part
15	50	35

3. At a summer sports camp, 105 children signed up for
soccer, 65 children signed up for softball, and 30 children
signed up for tennis. What fraction of the children chose
tennis?

Unit: _____   Whole: _____

Fraction: _____   Percent: _____

Total		
?		
Part	Part	Part
105	65	30

4. The students at Woodland Middle School have a bake
sale to raise money for the library. The sixth graders sell
600 items. The seventh graders sell 530 items. The eighth
graders sell 370 items. What fraction of the items do the
sixth graders sell?

Unit: _____   Whole: _____

Fraction: _____   Percent: _____

Total		
?		
Part	Part	Part
600	530	370

Use with or after Lesson 6•7.

Practice Set 42 continued

SRB
59-61

5. Describe a situation that the data in the line plot below might represent. Then give the plot a title and a unit.

Help find the landmarks,

Weight of 5 year old

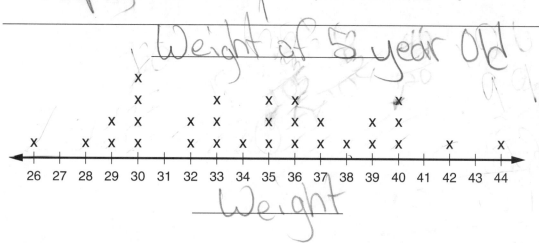

Weight

6. Find the following landmarks for the data in the line plot.

a. minimum: _26_   **b.** maximum: _44_   **c.** range: _18_

d. mode: _30_   **e.** median: _35_   **f.** mean: _35_

7. Writing/Reasoning How did you find the mean for the data?

I add all the numbers then divide it by 30 because there are 30 numbers

Use the division rule to write equivalent fractions.

8. $\frac{10}{12}$ = _____

9. $\frac{18}{72}$ = _____

10. $\frac{6}{30}$ = _____

11. $\frac{22}{24}$ = _____

12. $\frac{25}{200}$ = _____

13. $\frac{9}{45}$ = _____

14. $\frac{51}{120}$ = _____

15. $\frac{33}{99}$ = _____

16. $\frac{7}{420}$ = _____

17. $\frac{13}{52}$ = _____

18. $\frac{21}{51}$ = _____

19. $\frac{16}{48}$ = _____

Use with or after Lesson 6·7.

Practice Set 43

 Add or subtract.

1. $\frac{1}{3} + \frac{1}{3} =$ _____

2. $2\frac{1}{10} + \frac{3}{10} =$ _____

3. $\frac{7}{8} - \frac{2}{8} =$ _____

4. $3\frac{1}{2} + \frac{1}{2} =$ _____

5. $\frac{15}{16} - \frac{7}{16} =$ _____

6. $2\frac{5}{6} - \frac{11}{6} =$ _____

7. $\frac{2}{5} + \frac{2}{5} =$ _____

8. $9\frac{7}{12} - 2\frac{5}{12} =$ _____

 Solve.

9.
$$\begin{array}{r} 826 \\ -\ 104 \\ \hline \end{array}$$

10.
$$\begin{array}{r} 930 \\ -\ 285 \\ \hline \end{array}$$

11.
$$\begin{array}{r} 962 \\ +\ 268 \\ \hline \end{array}$$

12.
$$\begin{array}{r} 2,965 \\ -\ 1,583 \\ \hline \end{array}$$

13.
$$\begin{array}{r} 1,903 \\ -\ 825 \\ \hline \end{array}$$

14.
$$\begin{array}{r} 2,532 \\ +\ 7,378 \\ \hline \end{array}$$

15.
$$\begin{array}{r} 962 \\ +\ 25 \\ \hline \end{array}$$

16.
$$\begin{array}{r} 2,682 \\ -\ 632 \\ \hline \end{array}$$

17.
$$\begin{array}{r} 1,523 \\ +\ 1,497 \\ \hline \end{array}$$

18. Find the perimeter of each regular polygon.

Regular Polygon	Length of 1 side	Perimeter
square	15 cm	
pentagon	5.1 cm	
hexagon	$4\frac{1}{2}$ cm	
octagon	2.02 cm	

Use with or after Lesson 6•8.

Name _____ Date _____ Time _____

Practice Set 44

 Use clock fractions, if helpful, to solve these problems.
Write each answer as a fraction.

1. $\frac{5}{12} + \frac{5}{12} =$ _____

2. $\frac{1}{4} + \frac{1}{3} =$ _____

3. $\frac{5}{6} - \frac{1}{6} =$ _____

4. $\frac{11}{12} - \frac{1}{12} =$ _____

5. $\frac{3}{4} + \frac{1}{6} =$ _____

6. $\frac{5}{6} + \frac{1}{12} =$ _____

7. $\frac{2}{3} - \frac{1}{4} =$ _____

8. $\frac{5}{6} - \frac{1}{3} =$ _____

9. $\frac{1}{6} + \frac{1}{6} =$ _____

10. $\frac{13}{12} - \frac{5}{12} =$ _____

Write the number sentences with parentheses. Then solve.

11. Add 5.43 to the difference of 10.15 and 7.93.

$(5.43) + 2.22 = 7.65$ $(10.15 - 7.83) + 5.43 = 7.65$

12. Subtract the sum of 6 and 1.35 from 7.75.

$7.75 - 7.35 = 0.40$ $7.75 - (6 + 1.35) = 0.40$

2.22

13. Add 39 to the difference of 17.00 and 6.47.

$10.53 + 39 = 49.53$ $(17.00 - 6.47) + 39 = 49.53$

14. Subtract the sum of 81 and 8.92 from 848.37.

$848.37 - 72.08 = 776.29$ $848.37 - (81 + 8.92) =$

758.45

Solve.

15. How many 12s in 2,400? __200__

16. How many 70s in 8,400? __120__

17. How many 1,000s in 10^6? __1,000__

18. 16 * 80 = __1,280__

19. 84 * 50 = __4,200__

20. 600 * 8.3 = __4,980__

21. 29.3 * 9 = __263.7__

Use with or after Lesson 6·9.

Practice Set 44 continued

The pizza shown has been cut into 12 equal slices.

M = mushroom

P = pepperoni

S = sausage

O = onion

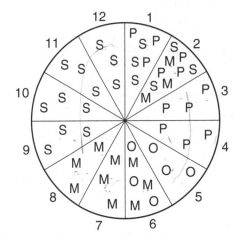

22. Write a decimal to show what part of the pizza has only
    one topping. __0.75__

23. What percent of the pizza has 2 or more toppings? __25__ %

24. What fraction of the slices has only sausage? __1/3__

25. What fraction of the pizza has no onions? __5/6__   10/12

26. If all the slices with mushrooms are eaten first, how many
    slices are left? __8__

27. What fraction of the remaining slices has pepperoni? __4/12 2/6 3/5__

28. What percent of the pizza has only mushrooms and/or onions?
    __33 1/3 %__

29. What fraction of the pizza has only pepperoni and/or sausage?
    __5/12__

30. Joe and Trish eat all the slices with sausage. What
    fraction of the pizza is left for Adam, Carrie, and Dave? __6/12__

31. If the three share the leftover pizza equally,
    how many slices should each one get? __2 1/2__

Use with or after Lesson 6·9.

Practice Set 45

Rewrite each pair of fractions as equivalent fractions with a common denominator.

1. $\frac{2}{3}, \frac{3}{4}$ $\frac{8}{12}$ $\frac{9}{12}$

2. $\frac{3}{5}, \frac{1}{2}$ $\frac{6}{10}$ $\frac{5}{10}$

3. $\frac{2}{5}, \frac{1}{3}$ $\frac{6}{15}$ $\frac{5}{15}$

4. $\frac{5}{6}, \frac{4}{5}$ $\frac{25}{30}$ $\frac{24}{30}$

5. $\frac{1}{2}, \frac{2}{3}$ $\frac{3}{6}$ $\frac{4}{6}$

6. $\frac{3}{7}, \frac{2}{3}$ $\frac{9}{21}$ $\frac{14}{21}$

7. $\frac{1}{3}, \frac{3}{10}$ $\frac{10}{30}$ $\frac{9}{30}$

8. $\frac{2}{5}, \frac{3}{7}$ $\frac{14}{35}$ $\frac{15}{35}$

9. $\frac{1}{4}, \frac{1}{5}$ $\frac{5}{20}$ $\frac{4}{20}$

10. $\frac{2}{7}, \frac{7}{8}$ $\frac{16}{56}$ $\frac{49}{56}$

Write each fraction as a percent.

11. $\frac{7}{10}$ 70%

12. $\frac{6}{100}$ 6%

13. $\frac{1}{2}$ 50%

14. $\frac{3}{4}$ 75%

15. $\frac{2}{5}$ 40%

16. $\frac{3}{20}$ 15%

Make a magnitude estimate for the quotient. Is the solution in the *tenths*, *ones*, *tens*, or *hundreds*?

17. $\frac{629}{9}$ tens

18. $\frac{32.1}{6}$ ones

19. $\frac{1.62}{7}$ tenths

20. $\frac{\$86.16}{5}$ tens

21. $\frac{678.1}{4}$ hundreds

22. $\frac{885}{21}$ tens

23. $\frac{239}{35}$ ones

24. $\frac{4.72}{6}$ tenths

25. **Writing/Reasoning** Describe a real-world situation in which you might need to estimate a quotient.

_____

_____

Use with or after Lesson 6·10.

Practice Set **45** *continued*

Complete.

26. $3^3 = $ _____

27. $4^{\square} = 64$

28. $8 * 8 * 8 * 8 = $ _____

29. 6 to the fourth power = _____

30. The square root of _____ = 11

31. The square root of 225 = _____

Rewrite the number sentences with parentheses to make them correct.

32. $8 * 10 - 5 = 75$ _____

33. $6 * 18 - 6 = 72$ _____

34. $42 - 21 - 7 = 14$ _____

35. $18 - 24 - 7 = 1$ _____

36. $5 * 8 + 2 * 13 = 650$ _____

37. $117 = 9 * 8 + 5$ _____

38. $236 = 4 * 8 + 51$ _____

39. $448 = 7 * 22 + 47 - 5$ _____

Solve.

40. How many 80s in 400? _____

41. How many 110s in 7,700? _____

 Solve.

42.
    12
  $* 2,100$

43.
    41
  $* 60$

44.
    119
  $* 3$

45.
  153
  $* 5$

46.
  286
  $* 23$

47.
  56
  $* 15$

48.
  820
  $* 16$

49. $26\overline{)364}$

50. $6\overline{)312}$

Use with or after Lesson 6•10.

Practice Set 46

1. Complete the table.

Exponential Notation	Base	Exponent	Repeated Factors	Standard Notation
2^6	2	6	2 * 2 * 2 * 2 * 2 * 2 *	64
	3	3		27
			8 * 8 * 8 * 8	
			5 * 5 * 5 * 5 * 5	
	7			5,764,801

2. **Writing/Reasoning** Which is true: $6^4 = 24$ or $6^4 = 1,296$?
Explain your answer.

_____

_____

Describe the mistake. Then find the correct solution.

3. $5^7 = 5 * 7 = 35$

Mistake:_____

Correct solution: _____

4. $3^5 = 3 + 5 = 8$

Mistake:_____

Correct solution: _____

Use your calculator to write the following numbers in standard notation.

5. $2^8 =$ _____

6. $7 * 7 * 7 * 7 * 7 =$ _____

7. $10 * 10 * 10 =$ _____

8. 9 to the 4th power = _____

9. $6 * 6 * 6 =$ _____

10. $4 * 4 * 4 * 4 * 4 * 4 * 4 * 4 * 4 =$ _____

11. $12^4 =$ _____

12. 1 to the 10th power = _____

Practice Set **46** *continued*

Write <, >, or =.

13. 6^7 _____ 7^6

14. 9^3 _____ 8^4

15. 5^5 _____ 3,125

16. 8^2 _____ 2^6

17. 3^6 _____ 6^5

18. 10^4 _____ 1,000

Find the missing angle measure.

19.

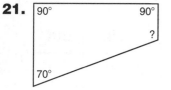

_____

20.

_____

21.

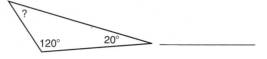

_____

22.

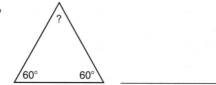

_____

Make a factor rainbow for each square number. Write the number, using exponents. Then complete the sentence.

23. 36

_____ = 36   The square root of 36 is _____.

24. 144

_____ = 144   The square root of 144 is _____.

25. 81

_____ = 81   The square root of 81 is _____.

Use with or after Lesson 7•1.

Practice Set 47

 Write the letter of the number that matches the expression.

1. $6 * 6 * 6$ _____   **A.** 15,625

2. $5 * 10^5$ _____   **B.** 8,000,000

3. 5^6 _____   **C.** 500,000

4. $3^2 * 4^3$ _____   **D.** 216

5. 3 to the seventh power _____   **E.** 576

6. $8 * 10^6$ _____   **F.** 2,187

Complete the "What's My Rule?" tables and rule boxes.

7.

Rule

out = in / 30

in	out
180	
240	
990	
1,260	
1,500	

8.

Rule

in	out
43	4
33	−6
	14
24	
71	32

9.

Rule

out = in * 105

in	out
6	630
8	
	945
11	
15	

10.

Rule

in	out
54	9
30	5
	6
1	
2	

Practice Set 47 continued

11. Use the clues to find the mystery number.

_____ _____ _____ , _____ _____ _____ , _____ _____ _____ , _____ _____ _____

- Find $\frac{2}{5}$ of 20. Write the result in the thousands place.

- Add 1 to the number in the thousands place. Write the answer in the ten-thousands place.

- Find 14 ∗ 4. Reverse the digits in the result and divide by 13. Write the result in the millions place.

- Add 6 to the digit in the ten-thousands place. Divide by 5 and write the result in the hundred-thousands place.

- Write $\frac{14}{2}$ as a whole number in the hundred-millions place.

- Find 60% of 10. Write the result in the ten-millions place.

- Subtract 3 from the number in the hundred-millions place. Write the result in the ten-billions place.

- Find $\frac{3}{7}$ of 49. Subtract 19 and write the result in the hundred-billions place.

- Find the sum of all the digits in the chart so far. Divide the result by 44 and write the answer in the billions place.

- Write 0 in the remaining places.

12. Write the number in words. _____

_____

Decide the chance of each event happening. Then write one of the following probability terms: *certain, unlikely,* or *50-50 chance.*

13. The sun will set tomorrow.   Chance: _____

14. A newborn baby will be a boy.   Chance: _____

15. It will snow at your house in July.   Chance: _____

16. February will have at least 28 days.   Chance:: _____

Practice Set 48

Write each number using scientific notation.

1. 6 million _____

2. 20,000 _____

3. 500,000 _____

4. 100 billion _____

5. 40,000,000 _____

6. 9 trillion _____

7. 3 thousand _____

8. 80 million _____

Write the numbers in order from least to greatest.

9. 1 million; $3 * 10^4$; 4 thousand; 100,000; $4 * 10^5$

_____

10. $6 * 10^5$; 2 million; 160,000; $5 * 10 * 10 * 10$; 25 thousand

_____

11. 33,000,000; $9 * 10^6$; 17 million; 1 billion; 4,000,000,000,000

_____

12. Ms. Ramsey's class collected information on students' favorite games. Complete the table and make a circle graph of the data. Remember to give your graph a title.

Favorite Games	Number of Students	Percent of Class
computer	10	
board	15	
card	5	
Total		

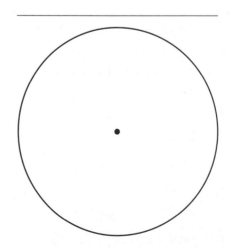

Practice Set 49

Rewrite the number models with parentheses to make them correct.

1. 28 / 4 − 7 = 0 _____

2. 23.2 = 8 ∗ 3 − 0.8 _____

3. 129 − 18 − 48 = 63 _____

4. 57.6 = 3 ∗ 12.8 + 6.4 _____

5. 11 ∗ 4.2 + 6 ∗ 10.1 = 106.8 _____

6. −70 = 7 ∗ 50 − 60 _____

7. 13 ∗ 14 + 8 − 3 = 187 _____

Write the letter of the prime factorization for each number.

8. 70 _____   **a.** 2 ∗ 3 ∗ 5   **b.** 2 ∗ 5 ∗ 7

9. 29 _____   **a.** 1 ∗ 29   **b.** 2 ∗ 5 ∗ 19

10. 80 _____   **a.** $2^5 ∗ 4$ **b.** $2^4 ∗ 5$

11. 28 _____   **a.** 2 ∗ 2 ∗ 7   **b.** $2^3 ∗ 72$

Write each fraction in simplest form.

12. $\frac{24}{4}$ _____   **13.** $\frac{36}{12}$ _____   **14.** $\frac{11}{11}$ _____   **15.** $\frac{70}{10}$ _____

16. $\frac{48}{8}$ _____   **17.** $\frac{18}{3}$ _____   **18.** $\frac{6}{4}$ _____   **19.** $\frac{8}{1}$ _____

20. $\frac{9}{12}$ _____   **21.** $\frac{40}{80}$ _____   **22.** $\frac{14}{16}$ _____   **23.** $\frac{24}{38}$ _____

24. $\frac{60}{30}$ _____   **25.** $\frac{25}{15}$ _____   **26.** $\frac{42}{14}$ _____   **27.** $\frac{20}{4}$ _____

Practice Set 50

Write *true* or *false* for each number sentence. Follow the rules for the order of operations.

1. (3 + 5) * 4 = 32 _____

2. (18 − 6) * 2 − 3 = 24 _____

3. (48 − 2²) ÷ 10 = 5 _____

4. 15 − 4 * 2 + 1 = 8 _____

5. 100 ÷ (25 + 25) + 25 = 27 _____

6. 16 − (8 + 2) = 10 _____

Write *acute, right, obtuse,* or *straight* to identify each angle.

7.

_____

8.

_____

9.

_____

10.

_____

COMPUTATION PRACTICE Solve. Write each answer as a mixed number.

11. 172 ÷ 8 = _____

12. 367 ÷ 12 = _____

13. 431 ÷ 6 = _____

14. 572 ÷ 3 = _____

15. 937 ÷ 3= _____

16. 795 ÷ 7 = _____

17. 538 ÷ 9 = _____

18. 641 ÷ 9 = _____

19. 373 ÷ 8 = _____

20. 823 ÷ 5 = _____

Practice Set 50 continued

Write the digit in the thousandths place.

21. 5.967 _____

22. 1.2350 _____

23. 8.84256 _____

24. 0.47000 _____

25. 3.368 _____

26. 10.96733 _____

Solve.

27. $360 - z = 241$ _____

28. $34 + 23 = y$ _____

29. $646 + 324 = n$ _____

30. $980 - 150 = x$ _____

31. $2,350 - m = 1,982$ _____

32. $90 = 10,800 / p$ _____

 Solve.

33. 478
      − 55

34. 363
      − 67

35. 34
      * 9

36. 3,436
      + 2,436

37. 3,456
      * 63

38. 16)384

39. An elephant can eat 500 pounds of hay and drink 60 gallons of water in one day. About how many pounds of hay would an elephant eat in a week?

_____

40. About how many pounds of hay does an elephant eat in a year?

_____

41. How many gallons of water does an elephant drink in one year?

_____

42. A box of pinwheel cookies contains 42 cookies. Richard and his six friends share the cookies equally. How many does each get?

_____

Use with or after Lesson 7•5.

Practice Set 51

Write the letter that identifies each number on the number line.

```
       C       A       G     D  H       B       F       E
   ◄—+—◆—+—◆—+—◆—+—◆—◆—+—◆—+—◆—+—◆—+—►
     −8 −7 −6 −5 −4 −3 −2 −1  0  1  2  3  4  5  6  7  8
```

1. −1 _____   **2.** 4 _____   **3.** 0 _____   **4.** 6 _____

5. −7 _____   **6.** 2 _____   **7.** −3 _____   **8.** −5 _____

Compare. Write <, >, or =.

9. −10 _____ $4\frac{1}{2}$ **10.** −3 _____ −5

11. −3.5 _____ $-3\frac{1}{2}$ **12.** −7 _____ 3

13. $-2\frac{1}{2}$ _____ $-2\frac{1}{4}$ **14.** $6\frac{1}{2}$ _____ $-3\frac{1}{3}$

15. −6.2 _____ $-6\frac{1}{5}$ **16.** 2.3 _____ 2.03

17. 5.0 _____ −0.5   **18.** $\frac{2}{3}$ _____ 0.6

Complete the number lines.

19.
```
   ◄—+——+——+——+——+——+—►
   8   ___  ___  ___  ___  38
```

20.
```
   ◄—+——+——+——+——+—►
  −2/4  ___  ___  ___  2/4
```

21.
```
   ◄—+——+——+——+——+——+——+—►
   3.4  ___  ___  ___  ___  4.6
```

22.
```
   ◄—+——+——+——+—►
   19   ___  ___  61
```

23.
```
   ◄—+——+——+——+——+—►
  −0.6  ___  ___  ___  0.2
```

Practice Set 51 continued

 Add or subtract.

24. $\frac{3}{10} + \frac{3}{10}$ _____

25. $\frac{11}{12} - \frac{1}{2}$ _____

26. $\frac{7}{8} - \frac{2}{8}$ _____

27. $\frac{3}{4} + \frac{1}{12}$ _____

28. $\frac{1}{4} + \frac{3}{4}$ _____

29. $\frac{5}{6} - \frac{1}{4}$ _____

30. $\frac{1}{2} + \frac{3}{8}$ _____

31. $\frac{3}{4} - \frac{1}{6}$ _____

Round 53,729,437 to the given place value.

32. hundred _____

33. thousand _____

34. hundred-thousand _____

35. million _____

Solve.

36. $60 * n = 3{,}600$ _____

37. $s * 7 = 49$ _____

38. $g * 32 = 640$ _____

39. $2 * m = 960$ _____

40. $b * 7 = 350$ _____

41. $j * 90 = 360$ _____

42. $8 * f = 640$ _____

43. $6 * a = 72$ _____

Use the clues to find the number.

44. Clue 1: I am a prime number less than 75.
Clue 2: My digits add up to 13.   _____

45. Clue 1: I am an even number.
Clue 2: I am less than 65, but greater than 50.
Clue 3: I am divisible by 5.   _____

46. Clue 1: I am an odd number.
Clue 2: I am divisible by 17.
Clue 3: I am less than 56, but greater than 34.   _____

Use with or after Lesson 7·6.

Practice Set 52

 Solve.

1. −15 + 2 = _____

2. −4 + −5 = _____

3. 62 + −9 = _____

4. −8 + −1 = _____

5. −14 + 6 = _____

6. 250 + −110 = _____

7. −61 + 60 = _____

8. −90 + −30 = _____

9. −43 + 43 = _____

10. 29 + −15 = _____

Use digits to write the following numbers.

11. one hundred twenty-two billion, three hundred twelve million, eighty-five thousand

_____

12. eighty-four and sixteen hundredths _____

13. eighteen trillion, two hundred thousand, fourteen

_____

Use words to write the following numbers.

14. 83,900,000,000,001 _____

_____

15. 14.657 _____

16. 4,296,087,050,000 _____

_____

Solve.

17. 18 + b = 142 _____

18. 900 ÷ c = 30 _____

19. 12 + t = 50 _____

20. n * 10 = 650 _____

21. a ÷ 5 = 70 _____

22. 16.75 − h = 12 _____

Practice Set 52 *continued*

For any pair of numbers on a number line, the number to the left
is less than the number to the right.

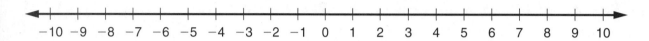

Write < or >.

23. −6 _____ 6

24. 12 _____ 8

25. −4 _____ 0

26. 10 _____ −1

27. $-3\frac{1}{2}$ _____ $-2\frac{1}{2}$

28. −1 _____ −3

Ms. Murphy's class made paper airplanes. Each group of students
tested their airplanes to see which stayed in the air the longest.
The table below shows the results of the test flights.

29. What is the minimum flight time reported by

the groups? _____

30. What is the range of the time the airplanes

were in the air? _____

31. Did Group A's or Group D's airplane fly longer?

_____

How much longer? _____

32. What is the combined flight time of all

5 groups? _____

Groups	Time in the Air
A	2.50 seconds
B	4.0 seconds
C	3.50 seconds
D	2.75 seconds
E	1.25 seconds

33. Plot the flight times from the table on the number line.
Label each time with the letter for the corresponding group.

Use with or after Lesson 7·7.

Practice Set 53

Use the thermometer number line to help you solve the subtraction problems.

> **Example** On Monday, the temperature was 8°F.
> By Tuesday, the temperature had dropped 15°F.
> What was the temperature on Tuesday?
>
> $\quad$ 8 $\quad$ Start at 8°F.
> -15 $\quad$ Go down 15°F.
> $\overline{\;-\;7}$ $\quad$ The result is −7°F, or 7° below zero Fahrenheit.

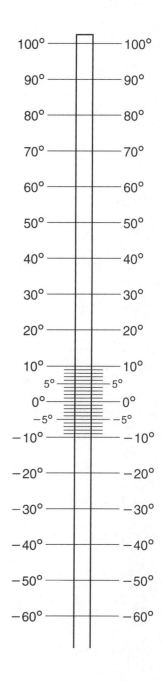

1. 80°F − 40°F = _____

2. 40°F − 60°F = _____

3. 6°F − 9°F = _____

4. 12°F − 18°F = _____

5. _____ = 80°F − 120°F

6. _____ = 60°F − 35°F

7. 25°F − 84°F = _____

8. 45°F − 39°F = _____

9. _____ = 6°F − 11°F

10. 29°F − 48°F = _____

11. 3°F − 25°F = _____

12. 32°F − 64°F = _____

13. −10°F − 45°F = _____

14. −90°F + 30°F = _____

15. −18°F + 72°F = _____

16. −21°F + −38°F = _____

Practice Set **53** *continued*

SRB
184 211

Write the next three numbers in the pattern.

17. 0.555, 0.535, 0.515, _____, _____, _____

18. 4.2, 3.2, 2.2, _____, _____, _____

19. 1, $\frac{1}{10}$, $\frac{1}{100}$, _____, _____, _____

20. −65, −85, −105, _____, _____, _____

Write the amounts.

21. 2 Ⓠ, 4 Ⓓ, 4 Ⓝ, 1 Ⓟ _____

22. 4 $1 , 1 Ⓠ, 2 Ⓓ, 1 Ⓝ, 2 Ⓟ _____

23. 3 $5 , 1 $1 , 2 Ⓠ, 1 Ⓝ _____

24. 1 $100 , 4 $20 , 1 $5 , 1 $1 , 1 Ⓠ, 14 Ⓝ, 17 Ⓟ _____

If 1 centimeter on a map represents 500 kilometers, then find the following:

> 1 km = 1000 m
> 1 m = 100 cm
> 1 cm = 10 mm

25. 7 cm represents _____ km

26. 280 mm represents _____ km

27. 63 cm represents _____ km

28. 9.5 cm represents _____ km

29. 65 mm represents _____ km

30. $\frac{1}{2}$ cm represents _____ km

31. $\frac{1}{5}$ cm represents _____ km

Use with or after Lesson 7·9.

Practice Set 54

SRB
91–94

Find the account balance.

1. 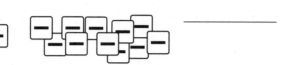 _____

2. 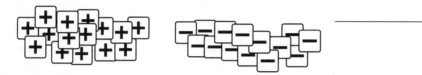 _____

3.  _____

4.  _____

5. 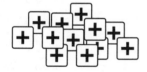 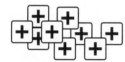 _____

6. **Writing/Reasoning.** Explain why these two tiles cancel each other out.

_____

_____

Practice Set 54 continued

Complete.

7. $100^2 =$ _____

8. $6^{\square} = 216$

9. $11 * 11 * 11 = 11^{\square}$

10. The square root of 676 = _____

 Solve.

11. $\dfrac{2}{8}$
$-\dfrac{1}{16}$

12. 48
$* 13$

13. 28
$* 39$

14. 79,430
$- 31,451$

15. 43
$* 92$

16. $\dfrac{6}{14}$
$-\dfrac{3}{4}$

17. 439
$+ 100$

18. 136,343
$* 5$

---

19. Write the missing numbers in the table.

Fraction	Percent	Decimal
$\dfrac{1}{5}$		
		0.57
	75%	
	81%	
		0.99

Use with or after Lesson 7•10.

Practice Set 55

Compare. Write <, >, or =.

1. $\frac{3}{5}$ _____ $\frac{10}{15}$

2. $\frac{3}{4}$ _____ $\frac{8}{12}$

3. $\frac{3}{10}$ _____ $\frac{2}{5}$

4. $\frac{3}{6}$ _____ $\frac{3}{7}$

5. $\frac{3}{9}$ _____ $\frac{1}{3}$

6. $\frac{5}{8}$ _____ $\frac{9}{16}$

7. $\frac{12}{20}$ _____ $\frac{2}{5}$

8. $\frac{7}{8}$ _____ $\frac{11}{12}$

9. $\frac{5}{6}$ _____ $\frac{2}{3}$

10. $\frac{1}{2}$ _____ $\frac{4}{7}$

11. $\frac{4}{5}$ _____ $\frac{8}{10}$

12. $\frac{7}{8}$ _____ $\frac{3}{4}$

Mr. Edwards records points for homework assignments.
Each student in the class is represented in the stem-and-leaf plot.

13. How many students are in the class?

    _____

14. What is the maximum?

    _____

15. What is the minimum?

    _____

16. What is the mode?

    _____

17. What is the median?

    _____

18. What is the mean?

    _____

19. What is the range?

    _____

Homework Points

Stems (10s)	Leaves (1s)
4	1 3 3 5 7
5	0 4 5 6 6 6
6	2 5 7 9 9
7	2 5 7 7
8	0

Practice Set 56

Rename each as a whole number or a mixed number in simplest form.

1. $\frac{12}{8}$ _____

2. $\frac{6}{3}$ _____

3. $\frac{27}{8}$ _____

4. $\frac{16}{5}$ _____

5. $\frac{36}{4}$ _____

6. $\frac{19}{6}$ _____

 Write each sum as a whole number or mixed number in simplest form.

7. $3\frac{1}{2} + 2\frac{1}{2} =$ _____

8. $4\frac{1}{8} + 2\frac{3}{8} =$ _____

9. $7\frac{2}{9} + 3\frac{8}{9} =$ _____

10. $2\frac{3}{10} + 4\frac{1}{5} =$ _____

11. $3\frac{1}{6} + 4\frac{5}{12} =$ _____

12. $4\frac{1}{12} + 1\frac{1}{3} =$ _____

Tell whether each number sentence is *true* or *false*.

13. $20 + (10 * 7.4) = 94$

_____

14. $70 = (11 * 5) + 18$

_____

15. $32 + (18 * 6) = 132$

_____

16. $35 + (\frac{99}{9}) = 24$

_____

17. **Writing/Reasoning** How can adding parentheses change the answer to an equation? Give one example to support your answer.

_____

_____

Order these numbers from least to greatest.

18. 4.8   0.84   4.008   4,000.08   80,000.4

_____

19. 20.1   2.01   0.21   201.2   120.1

_____

Use with or after Lesson 8•2.

Practice Set 56 *continued*

> The three angles of a triangle always add up to 180°. A right angle measures 90°.

20. What is the measure of angle *T*?

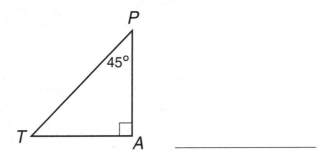

_____

21. What is the measure of angle *R*?

_____

Fill in the missing numbers on the number lines.

22.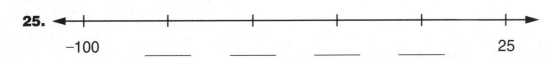

    − 18   _____   _____   − 9   _____   − 3

23.

    106   _____   _____   _____   656

24.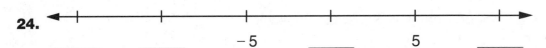

    _____   _____   − 5   _____   5   _____

25.

    −100   _____   _____   _____   _____   25

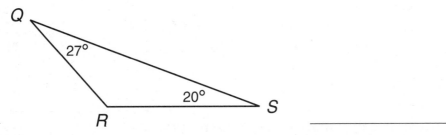

Practice Set 57

SRB
71 72
222 223

 Subtract. Write your answers in simplest form.

1. $5 - \frac{5}{6} =$ _____

2. $9 - \frac{5}{12} =$ _____

3. $6 - 1\frac{1}{4} =$ _____

4. $10 - 8\frac{2}{3} =$ _____

5. $3\frac{3}{4} - 2\frac{1}{4} =$ _____

6. $9\frac{5}{6} - 7\frac{1}{6} =$ _____

7. $9\frac{1}{5} - 4\frac{2}{5} =$ _____

8. $7\frac{11}{12} - 2\frac{5}{12} =$ _____

 Solve.

9. $(15 - 15) * 4 =$ _____

10. $65 - (6 * 9) =$ _____

11. $(560 + 70) / 30 =$ _____

12. $(900 / 3) + 40 =$ _____

13. The temperature at noon is 18°F. With the wind chill, it feels like −4°F. How much colder is the wind chill temperature than the actual temperature? _____

Add parentheses to the number sentences to make them correct.

14. $880 = 80 * 16 + 2 - 7$

15. $7 * 12 - 6 = 42$

16. $144 = 8 + 4 * 12$

17. $37 - 18 - 5 = 24$

18. $45 - 18 + 31 = -4$

19. $56 = 8 + 4 * 12$

20. $45 - 18 + 31 = 58$

21. $5 * 8 + 6 * 9 = 94$

Solve.

22. You start with a number. Double it. Square the answer. You get 400. What number did you start with?

_____

23. You start with a number. Double it. Square the answer. You get 1,296. What number did you start with?

_____

Use with or after Lesson 8+3.

Practice Set **57** *continued*

Write <, >, or = to make each sentence true.

24. $\frac{1}{4} + \frac{1}{2}$ _____ $\frac{4}{6}$

25. $2\frac{4}{12}$ _____ 2.5

26. $\frac{12}{2}$ _____ $\frac{80}{10}$

27. 12.5 _____ $\frac{25}{2}$

28. $\frac{3}{5}$ _____ $\frac{4}{10}$

29. $3\frac{3}{8}$ _____ $\frac{25}{8}$

Complete the "What's My Rule?" tables.

30.

Rule
out = in / 3

in	out
4,260	1,420
2,100	
6,399	
4,584	
	85

31.

Rule
out = in $*$ 3.5

in	out
27	
33	
	14
24	
51	

Solve.

32. $-7 + x = 0$

$x =$ _____

33. $-12 - y = 0$

$y =$ _____

34. $5 * t = 1$

$t =$ _____

35. $\frac{1}{8} * r = 1$

$r =$ _____

36. $r - 6 = 15$

$r =$ _____

37. $\frac{1}{11} * t = 3$

$t =$ _____

38. $12x = 144$

$x =$ _____

39. $4 + y = 18$

$y =$ _____

40. $x + 5 = -10$

$x =$ _____

Practice Set 58

Write the letter of the picture that best represents each expression.

1. $\frac{1}{4}$ of $\frac{1}{2}$ _____

A.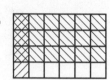

2. $\frac{2}{3}$ of $\frac{1}{4}$ _____

B.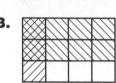

3. $\frac{1}{2}$ of $\frac{5}{6}$ _____

C.

4. $\frac{3}{4}$ of $\frac{1}{6}$ _____

D.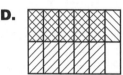

---

5. What is the name of the polygon?

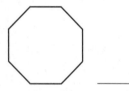 _____

6. If each side is 6.5 inches, what is the perimeter of the figure?

_____

7. How many lines of symmetry does this figure have?

_____

8. **Writing/Reasoning** Draw a shape that will look the same when it is rotated 180° clockwise. Explain why.

_____

_____

Use with or after Lesson 8·5.

Practice Set 59

Use area models to complete the problems.

Example

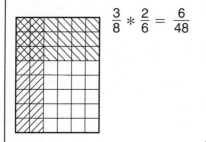

$$\frac{3}{8} * \frac{2}{6} = \frac{6}{48}$$

1.

$$\frac{1}{4} * \frac{1}{5} = \underline{\hspace{2cm}}$$

2.

$$\frac{3}{4} * \frac{1}{3} = \underline{\hspace{2cm}}$$

3.

$$\frac{5}{8} * \frac{4}{7} = \underline{\hspace{2cm}}$$

4.

$$\frac{4}{5} * \frac{5}{6} = \underline{\hspace{2cm}}$$

5.

$$\frac{2}{3} * \frac{3}{5} = \underline{\hspace{2cm}}$$

COMPUTATION PRACTICE **Solve.**

6. $\frac{1}{3} * \frac{3}{5} = \underline{\hspace{2cm}}$

7. $\frac{2}{9} * \frac{1}{6} = \underline{\hspace{2cm}}$

8. $\frac{1}{4} * \frac{4}{7} = \underline{\hspace{2cm}}$

9. $\frac{1}{2} * \frac{4}{15} = \underline{\hspace{2cm}}$

10. $\frac{7}{10} * \frac{5}{10} = \underline{\hspace{2cm}}$

11. $\frac{4}{5} * \frac{5}{9} = \underline{\hspace{2cm}}$

Complete the equivalent fractions.

12. $\frac{1}{6} = \frac{\square}{30}$

13. $\frac{2}{\square} = \frac{1}{7}$

14. $\frac{6}{8} = \frac{\square}{20}$

15. $\frac{9}{\square} = \frac{18}{50}$

Use the division rule to find equivalent fractions.

16. $\frac{25}{35} = \underline{\hspace{1.5cm}}$

17. $\frac{6}{9} = \underline{\hspace{1.5cm}}$

18. $\frac{18}{30} = \underline{\hspace{1.5cm}}$

19. $\frac{12}{60} = \underline{\hspace{1.5cm}}$

Practice Set 60

 Find the product. Use area models to help you.

1. $5 * \frac{2}{3} =$ _____

2. $\frac{1}{5} * 6 =$ _____

3. $4 * \frac{1}{4} =$ _____

4. $7 * \frac{1}{3} =$ _____

5. $\frac{3}{5} * 6 =$ _____

6. $4 * \frac{3}{8} =$ _____

Compare. Write < or >.

7. -3.8 _____ -2

8. -1.03 _____ -1.3

9. 0.5 _____ -1.5

10. 23 _____ -25

11. -45 _____ -52

12. -1.0 _____ -0.1

 Write a number sentence, and then solve. Tell what you did about the remainder.

13. Mari sells eggs by the dozen. One week her chickens laid 152 eggs. How many dozen eggs did she have to sell that week?

_____

_____

14. There are 137 fifth graders going on a field trip to the museum. The school buses each hold 62 students. How many buses are needed for the trip?

_____

_____

15. Sheri weighs a bag of tomatoes. It weighs 40 ounces. How many pounds of tomatoes are in the bag?

_____

_____

16. Vaughn has a collection of 13 miniature classic cars. Each display shelf holds 5 cars. How many shelves does Vaughn need to display his entire collection?

_____

_____

Use with or after Lesson 8·7.

Practice Set 60 *continued*

Use digits to write the following numbers.

17. six hundred eighty-six and thirty-eight hundredths

_____

18. three billion, four hundred fourteen million, six hundred
ninety-one thousand

_____

19. nine hundred sixty-eight and eleven thousandths

_____

20. six trillion, seventy-two billion, eighteen

_____

Use words to write the following numbers.

21. 2,000,000,002 _____

22. 32.906 _____

23. 0.505 _____

Complete the "What's My Rule?" tables.

24.

Rule		in	out
		7	$15\frac{1}{4}$
		10	$18\frac{1}{4}$
			20
		$13\frac{1}{2}$	
		$22\frac{3}{4}$	31

25.

Rule	in	out
out = in $*$ 7	4	
		35
		70
	6	
	8	

Practice Set 61

COMPUTATION PRACTICE **Multiply. Write your answer in simplest form.**

1. $3\frac{1}{3} * \frac{4}{5} =$ _____

2. $\frac{1}{2} * \frac{5}{8} =$ _____

3. $7 * 2\frac{1}{3} =$ _____

4. $4\frac{1}{6} * 2\frac{2}{5} =$ _____

5. $3\frac{1}{8} * 6\frac{3}{5} =$ _____

6. $3\frac{1}{3} * \frac{3}{10} =$ _____

7. $6\frac{3}{5} * 4 =$ _____

8. $5\frac{1}{7} * 8 =$ _____

9. $3\frac{2}{5} * 7\frac{2}{3} =$ _____

10. $2\frac{2}{9} * 9\frac{3}{8} =$ _____

Complete the following number lines.

11.

500   6,500

12.

15   355

13.

2   82

Round each number to the nearest hundredth.

14. 18.985 _____

15. 5.264 _____

16. 344.745 _____

17. 23.536 _____

18. 29.904 _____

19. 0.026 _____

20. 461.066 _____

21. 0.178 _____

22. 82.721 _____

23. 7.921 _____

24. 94.387 _____

25. 102.431 _____

26. 521.767 _____

27. 187.889 _____

Use with or after Lesson 8·8.

Practice Set 62

Write a decimal and a percent for each fraction.

1. $\frac{73}{100}$ _____

2. $\frac{1}{2}$ _____

3. $\frac{3}{4}$ _____

4. $\frac{45}{100}$ _____

5. $\frac{1}{3}$ _____

6. $\frac{3}{5}$ _____

7. $\frac{9}{10}$ _____

8. $\frac{8}{50}$ _____

9. $\frac{3}{20}$ _____

Maury bought a new shirt that was on sale at 15% off the original price. The original price was $30.

10. How much did Maury pay for the shirt? _____

11. How much money did Maury save on the shirt? _____

Measure to the nearest $\frac{1}{2}$ cm.

12. ├───────┤   _____ cm

13. ├─────────────────┤   _____ cm

 Solve.

14. 12
   $*\ 9$

15. 52
   $-\ 38$

16. 258
  $+\ 3{,}217$

17. 6,323
   $-\ 236$

18. 10
   $*\ 8$

19. 75
   $-\ 22$

20. 876
  $+\ 921$

21. 8,614
   $+\ 125$

22. $1\frac{2}{3}$
   $-1\frac{1}{3}$

23. $7\frac{5}{8}$
   $+1\frac{1}{8}$

24. $9\frac{7}{10}$
   $+3\frac{3}{10}$

25. $6\frac{2}{5}$
   $+4\frac{2}{5}$

Practice Set 62 *continued*

Sally, John, and Jeff drove from Denver to Chicago.
Sally drove $\frac{2}{9}$ of the distance. John drove $\frac{1}{3}$ of the
distance. Jeff drove the rest of the way.

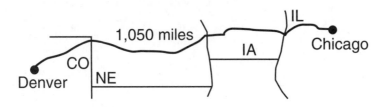

26. How many miles did Sally drive? _____

27. How many miles did John drive? _____

28. How many miles did Jeff drive? _____

29. **Writing/Reasoning** Explain how you found the answer
to Problem 28.

_____

_____

_____

Write <, > , or = to make each sentence true.

30. $\frac{1}{4}$ _____ $\frac{3}{6}$ **31.** $1\frac{4}{12}$ _____ $\frac{5}{3}$ **32.** $\frac{8}{2}$ _____ $\frac{9}{10}$

33. 4.5 _____ $\frac{25}{5}$ **34.** $\frac{1}{5}$ _____ $\frac{2}{10}$ **35.** 2.25 _____ $\frac{30}{10}$

Complete the "What's My Rule?" tables.

36.

Rule

out = in ∗ 710

in	out
6	4,260
8	
9	
11	
15	

37.

Rule

out = in / 3

in	out
27	
33	
	14
24	
51	

Use with or after Lesson 8•9.

Practice Set 63

 Solve.

1. $\frac{1}{10}$ of 90 _____

2. $\frac{2}{5}$ of 250 _____

3. $\frac{3}{8}$ of 24 _____

4. $\frac{1}{40}$ of 160 _____

5. $\frac{9}{16}$ of 32 _____

6. $\frac{2}{4}$ of 90 _____

7. $\frac{9}{8}$ of 64 _____

8. $\frac{9}{9}$ of 81 _____

9. $\frac{7}{8}$ of 16 _____

10. $\frac{6}{8}$ of 72 _____

11. $\frac{4}{3}$ of 18 _____

12. $\frac{5}{6}$ of 30 _____

---

13. **Use the clues to write a seven-digit number.**

_____ , _____ _____ _____ , _____ _____ _____

• Multiply 9 by 12. Subtract 100. Write the result in the hundreds place.

• Triple the number in the hundreds place and then divide by 4. Write the result in the millions place.

• Divide 3,300 by 1,100. Add 1 and write the result in the hundred-thousands place.

• Double the number in the millions place and divide by 6. Write the result in the tens place.

• Add 3 to the number in the tens place. Write the result in the ten-thousands place.

• Find 2% of 50. Write the result in the ones place.

• Divide 630 by 90. Write the result in the thousands place.

• Write the mystery number in words. _____

Use with or after Lesson 8·10.

Practice Set 63 continued

SRB
73-75
243-245

Solve.

14. 18 is $\frac{2}{3}$ of what number? _____

15. $\frac{4}{6}$ is $\frac{1}{9}$ of what number? _____

16. 12 is 80% of what number? _____

17. 52 is 13% of what number? _____

18. At Dan's school, $\frac{5}{7}$ of the students buy the school lunch. If 210 students buy the school lunch, how many students are there altogether?

_____

19. Alaina took $\frac{11}{13}$ of her stamp collection to school to show her classmates. If she took 143 stamps to school, how many total stamps does Alaina have in her collection?

_____

20. The community baseball team played 12 games. Each game was 7 innings long. Rylan pitched $\frac{5}{12}$ of all the innings the baseball team played. How many innings did Rylan pitch?

_____

There are 32 backpacks lined up against the gym wall.
One-half of the backpacks are blue. One-fourth of the backpacks
are red. One-eighth of the backpacks are green. Four of the
backpacks are black.

21. How many backpacks are blue? _____

22. How many backpacks are red? _____

23. How many backpacks are green? _____

Complete the tables.

24.

$\frac{1}{4}$ of 36	
$\frac{1}{2}$ of 36	
$\frac{3}{4}$ of 36	
$\frac{4}{4}$ of 36	

25.

25% of 48	
50% of 48	
75% of 48	
100% of 48	

Use with or after Lesson 8•10.

Practice Set 64

Solve.

Graph A

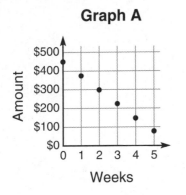

Amount / Weeks

Graph B

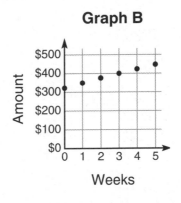

Amount / Weeks

Graph C

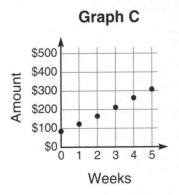

Amount / Weeks

1. Susan deposited $45 every week during a five-week period. After three weeks, she had a little more than $200. Which graph depicts her account value?

_____

2. If Susan continued making regular weekly deposits, about how much money would she have in her account after nine weeks?

_____

3. Tom withdrew $75 every week for five weeks. Which graph depicts his account value?

_____

4. If Tom continued making regular withdrawals, at what point will Tom run out of money?

_____

5. After five weeks of regular deposits, Julian's account balance was $450. Which graph depicts his account value?

_____

6. If Julian continued making regular weekly deposits, how long would it take him to save $1,000?

_____

Practice Set **64** *continued*

 Solve.

7.   300
      − 600

8. 62,473
     + 5,268

9.   253
      + 253

10. 2,352
      − 967

11. 2,675
      + 1,006

12.   35
       − 63

13.   264
       + 658

14. 12,965
      − 1,583

15. 4,322
      − 3,362

16. How much is $\frac{5}{8}$ of 32¢? _____

17. How much is $\frac{2}{12}$ of 54¢? _____

18. How much is $\frac{1}{10}$ of $8.30? _____

19. How much is $\frac{1}{3}$ of $3.60? _____

20. How much is $\frac{2}{5}$ of $2.20? _____

21. How much is $\frac{6}{3}$ of 27¢? _____

Complete.

22. $3^4 =$ _____

23. $5^{\square} = 3,125$

24. $6 * 6 * 6 * 6 =$ _____

25. 10 to the fourth power = _____

**Rewrite the number sentences with parentheses
to make them correct.**

26. $43 - 24 - 8 = 27$ _____

27. $19 - 35 - 8 = -8$ _____

28. $6 * 9 + 3 * 14 = 1,008$ _____

29. $240 = 10 * 6 + 18$ _____

30. $370 = 5 * 9 + 65$ _____

Use with or after Lesson 9·1.

Name _____ Date _____ Time _____

Practice Set 65

Write the ordered pair for each point.

1. A (2,3)
2. C (-3,-1)
3. E (-2,2)
4. G (1,-3)

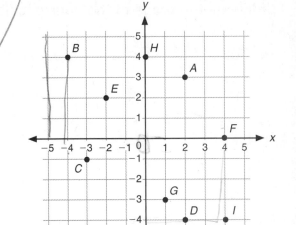

Name the point for each ordered pair.

5. (0,4) H
6. (-4,4) B
7. (4,-4) I
8. (2,-4) D

Write the next three numbers in each pattern.

9. 40, 80, 160, __320__, __640__, __1280__

10. 9, 18, 36, __72__, __144__, __288__

11. $\frac{6}{4}, \frac{12}{4}, \frac{24}{4}$, $\frac{48}{4}$, $\frac{96}{4}$, $\frac{192}{4}$

12. $\frac{1}{8}, \frac{3}{8}, \frac{5}{8}$, $\frac{7}{8}$, $\frac{9}{8}$, $\frac{11}{8}$

13. -0.8, -0.3, 0.2, __0.7__, __1.2__, __1.7__

14. 6.5, 3.25, 0, __-3.25__, __-6.50__, __-9.75__

Write the amounts.

15. Q Q Q D D N N P P P $1.08

16. $1 Q Q Q D D N N N P P $2.12

17. $1 $1 $1 $1 $1 $1 $1
Q N $7.30

Practice Set 66

Calculate the areas of the figures below.

1. Scale: Each square = 1 square inch

Area = ___14___ in²

2. Scale: Each square = 1 square foot

Area = ___53___ ft²

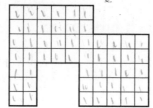

3. Scale: Each square = 1 square centimeter

Area = ___28___ cm²

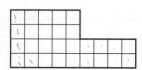

4. Scale: Each square = 1 square meter

Area = ___36___ m²

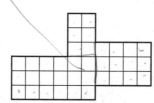

5. Scale: Each square = 1 square meter

Area of triangle ABC = ___15___ m²

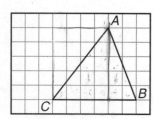

6. Scale: Each square = 1 square inch

Area of triangle DEF = ___17___ in²

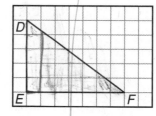

7. Scale: Each square = 1 square foot

Area of triangle HIJ = ___14___ ft²

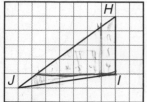

8. Scale: Each square = 1 square mile

Area of KLMN = ___24___ square miles

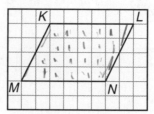

120

Practice Set 66 *continued*

Plot and label the point for each ordered pair on the coordinate grid.

9. G (7,7)   **10.** H (9,8)

11. I (0,10)   **12.** J (10,0)

13. K (9,9)   **14.** L (0,0)

Write the ordered pair for each point on the coordinate grid.

15. A (_____, _____)   **16.** B (_____, _____)

17. C (_____, _____)   **18.** D (_____, _____)

19. E (_____, _____)   **20.** F (_____, _____)

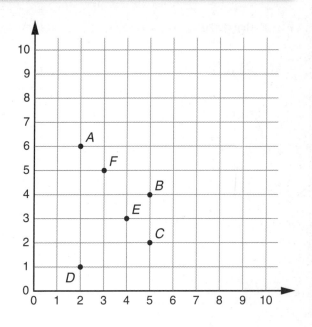

Complete.

21. Draw $\overline{AF}$. Draw $\overline{EC}$. What is the relationship between $\overline{AF}$ and $\overline{EC}$?

_____

22. Draw $\overline{AD}$. What kind of angle is angle *DAF*? _____

23. Draw $\overline{BD}$. What is the relationship between $\overline{BD}$ and $\overline{EC}$?

_____

24. Connect points *A, B, C,* and *D.* What type of polygon is *ABCD*? _____

25. Name the two parallel sides in polygon *ABCD.* _____

26. What is the sum of the angles in polygon *ABCD*? _____

27. **Writing/Reasoning** Explain how you found the answer to Problem 26.

_____

_____

Practice Set 67

Find the area.

1.

3.5 m

5.5 m

_____

2.

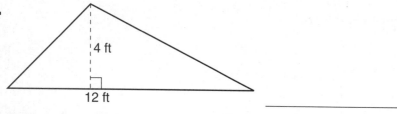

4 ft

12 ft

_____

3.

2.6 km

3.8 km

_____

4. ✏️ **Writing/Reasoning** Describe how this shape would look after it is translated.

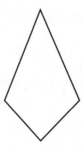

_____

_____

_____

Use the following list of numbers to answer the questions.

18.5, 16.25, 15.75, 13.5, 19.25, 11.5, 22.5, 14.25, 11.5

5. What is the range? _____

6. What is the mode? _____

7. What is the median? _____

8. What is the mean? _____

Practice Set 67 *continued*

 Solve.

9. 375
   * 24

10. 65.9
   + 93.6

11. 41.70
   + 2.57

12. 3.88
   − 2.92

13. 12
   * 18

14. 63
   − 73

15. $\frac{2}{3}$
   $+ \frac{4}{3}$

16. $\frac{14}{8}$
   $- \frac{1}{16}$

17. 634
   + 274

18. 3,435
   + 285

19. 28
   * 9

20. 378
   * 5

Complete the "What's My Rule?" tables.

21.

Rule
out = in + $\frac{1}{4}$

in	out
$\frac{1}{4}$	
$\frac{1}{2}$	
$\frac{3}{4}$	
$\frac{2}{8}$	
$\frac{5}{8}$	

22.

Rule
out = in − 1.1

in	out
2.7	
3.3	
	1.4
2.4	
8	6.9

Complete.

23. 84 days = _____ weeks

24. 16 weeks = _____ months

25. 2.5 hours = _____ seconds

26. 360 minutes = _____ hours

27. 10% of a day = _____ minutes

28. 180 seconds = _____ minutes

Practice Set 68

1. Write the formula for the volume of a rectangular prism.

_____

What does each letter in the formula represent?

_____

Find the volume of each rectangular prism below.

2.
6 ft
6 ft
5 ft

$V =$ _____ ft³

3.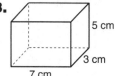
5 cm
3 cm
7 cm

$V =$ _____ cm³

4.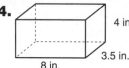
4 in.
3.5 in.
8 in.

$V =$ _____ in³

5.
8 ft
3 ft
6 ft

$V =$ _____ ft³

6.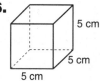
5 cm
5 cm
5 cm

$V =$ _____ cm³

7.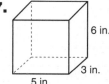
6 in.
3 in.
5 in.

$V =$ _____ in³

8.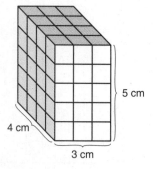
5 cm
4 cm
3 cm

Area of base = _____ cm²

Volume of first layer = _____ cm³

Volume of entire cube structure = _____ cm³

9.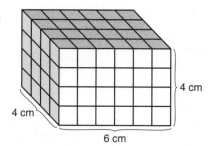
4 cm
4 cm
6 cm

Area of base = _____ cm²

Volume of first layer = _____ cm³

Volume of entire cube structure = _____ cm³

Use the division rule to find equivalent fractions.

10. $\frac{3}{9} =$ _____

11. $\frac{24}{60} =$ _____

12. $\frac{27}{30} =$ _____

13. $\frac{9}{21} =$ _____

Use with or after Lesson 9·8.

Practice Set 69

SRB
195–197

Find the volume of each prism. Include the units.

Volume = length * width * height
Volume = area of base * height

1.

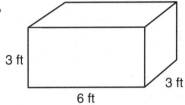

3 ft

6 ft

3 ft

V = _____

2.

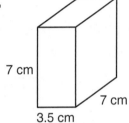

7 cm

7 cm

3.5 cm

V = _____

3.

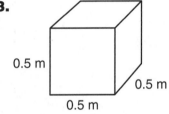

0.5 m

0.5 m

0.5 m

V = _____

4.

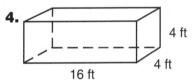

4 ft

16 ft

4 ft

V = _____

5.

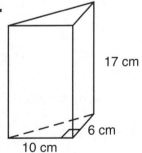

17 cm

6 cm

10 cm

V = _____

6.

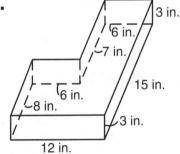

3 in.

6 in.

7 in.

6 in.

8 in.

15 in.

3 in.

12 in.

V = _____

Practice Set 69 continued

Write the next three numbers in each pattern.

7. −15, −10, −5, _____, _____, _____

8. 0.04, 0.06, 0.08, _____, _____, _____

9. 0.44, 0.68, 0.92, _____, _____, _____

**Use the information in the table to answer the questions below.
Round to the nearest whole number.**

Year	U.S. Population	Number of Children Ages 5 to 14
1900	76,000,000	17,000,000
1990	255,000,000	28,000,000

10. In 1900, about what percent of the population was 5 to 14 years old? _____

11. About what percent of the population was 5 to 14 years old in 1990? _____

12. About how many times larger was the entire U.S. population in 1990

than in 1900? _____

13. About how many times larger was the population of 5- to 14-year olds?

_____

14. Did the ratio of children to the U.S. population increase or decrease

from 1900 to 1990? _____

15. **Writing/Reasoning** Why do you think the ratio in Problem 14
changed? Explain your answer.

_____

_____

16. 76,000,000 can be written using scientific notation as 7.6×10^7.
Write all other numbers in the table, using scientific notation.

_____

Use with or after Lesson 9•9.

Practice Set 70

Complete.

1. 3 L = _____ mL

2. 150 cm³ = _____ mL

3. 0.5 L = _____ mL

4. 4,200 cm³ = _____ L

5. 1 gal = _____ qt

6. 3 qt = _____ pt

7. 96 oz = _____ qt

8. 3 c = _____ oz

---

9. Find the area of each rectangle below. Write a number model to represent each.

Example   Rectangle *A*: 2 ∗ 5 = 10 square units

height (or width)

base (or length)

B

A

C

F

D

E

_____

_____

_____

_____

Practice Set 71

Solve the pan-balance problems.

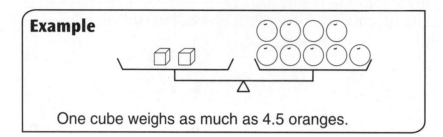

Example

One cube weighs as much as 4.5 oranges.

1.

One orange weighs as much as _____ grapes.

2.

One block weighs as much as _____ marbles.

3.

One block weighs as much as _____ marbles.

4.

One cube weighs as much as _____ balls.

Use with or after Lesson 10·1.

Practice Set 72

In each figure below, the two pans are in perfect balance. Solve these pan-balance problems, using both pan balances. The weights of objects, such as blocks, marbles, balls, and coins, are consistent within each problem.

1. One block weighs as much as _____ marbles.

One ball weighs as much as _____ marbles.

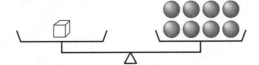

2. *x* weighs as much as _____ marbles.

y weighs as much as _____ marbles.

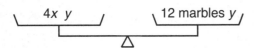

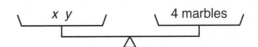

3. **Writing/Reasoning** Explain how you solved Problem 2.

_____

_____

_____

COMPUTATION PRACTICE Write the product as a fraction.

4. $\frac{1}{2}$ of $\frac{1}{4}$ = _____

5. $\frac{1}{8}$ of $\frac{1}{4}$ = _____

6. $\frac{1}{2}$ of $\frac{1}{2}$ = _____

7. $\frac{1}{3}$ of $\frac{1}{6}$ = _____

8. $\frac{1}{4}$ of $\frac{1}{4}$ = _____

9. $\frac{1}{4}$ of $\frac{1}{12}$ = _____

COMPUTATION PRACTICE Solve.

10. 1% of 87 = _____

11. 19% of 85 = _____

12. 35% of 628 = _____

13. 25% of 44 = _____

14. 75% of 150 = _____

15. 9% of 48 = _____

Practice Set 72 continued

 Solve.

16. 2.5 * 4.3 = _____

17. 1.2 * 0.35 = _____

18. 41.7 * 0.8 = _____

19. 58.285 + 18.572 = _____

20. 845.9 + 38.2 = _____

21. 582.967 + 93.837 = _____

22. 306.403 − 217.284 = _____

23. 572.25 − 23.41 = _____

24. For relay races, the gym teacher divided the class into 5 teams with an equal number of students on each team. Extra students helped judge the race. There were 28 students. How many judges were there? _____

25. Ms. Krupa made 146 ounces of stewed tomatoes. How many 6-ounce jars can she fill? _____

Simplify the fractions.

26. $1\frac{8}{20}$ _____

27. $4\frac{8}{12}$ _____

28. $\frac{54}{6}$ _____

29. $8\frac{12}{48}$ _____

30. $\frac{26}{12}$ _____

31. $\frac{19}{4}$ _____

32. $\frac{90}{15}$ _____

33. $\frac{22}{11}$ _____

34. $\frac{15}{45}$ _____

35. $\frac{27}{81}$ _____

36. $\frac{42}{77}$ _____

37. $\frac{16}{56}$ _____

Write number sentences for the following. Then tell whether each is *true* or *false*.

38. If 8.5 is subtracted from 24.6, the result is 16.1. _____

39. 11 is twice as much as 5.5. _____

40. 275 is more than 700 * $\frac{3}{4}$. _____

41. Divide 68 by 2 and you get 34. _____

42. If 58 is decreased by 16, the result is −48. _____

43. 45 is greater than the sum of 76 and (−35). _____

Practice Set 73

Complete the "What's My Rule?" tables and the rule box.

1.

Rule
out = in * 14

in	out
3	
5	
7	
11	
15	

2.

Rule
out = in / 15

in	out
30	
105	
705	
1,170	
1,530	

3.

Rule
out = in − 75

in	out
60	
123	
	35
11	
	−82

4.

Rule

in	out
8	18.25
14	24.25
	44.75
35	
48.75	59

Find the following landmarks for the set of numbers below.

27, 18, 46, 33, 30, 27, 14, 25

5. maximum _____

6. minimum _____

7. range _____

8. median _____

9. mean _____

10. mode _____

Compare. Write <, >, or =.

11. −3.05 _____ −3.15

12. −127 _____ −172

13. $\frac{10}{12}$ _____ $\frac{5}{6}$

14. $1\frac{1}{5}$ _____ $\frac{7}{5}$

15. −1.5 _____ $-1\frac{1}{2}$

16. −0.42 _____ $-\frac{6}{10}$

Practice Set 74

SRB
45 46
233 234

Complete each table below. Then graph the data and connect the points.

1. **a.** The bakery is selling muffins for $3.50 per dozen. Rule: Cost = $3.50 ∗ number of dozen

Dozens (d)	Cost ($) = (3.50 ∗ d)
1	
3	
5	
	24.50

b. Plot a point to show the cost of 9 dozen muffins. How much would 9 dozen muffins cost? _____

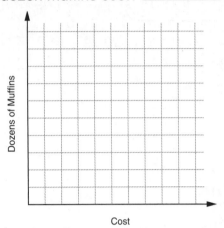

Cost

2. **a.** The weather report says that snow is falling at the rate of 1.75 inches per hour. Rule: Inches of snow = 1.75 in. ∗ number of hours

Hours (h)	Inches of Snow (1.75 ∗ h)
1	
3	
	10.5
9	

b. Plot a point to show how much snow will fall in 12 hours. How much snow will fall in 12 hours? _____

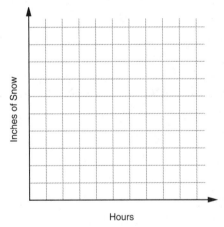

Hours

3. **Writing/Reasoning** Using the graph, the rule, or the table, explain how to find the amount of snow that will fall in 5.5 hours.

_____

_____

_____

Round the decimals to the nearest tenth.

4. 23.87 _____

5. 108.43 _____

6. 389.21 _____

7. 19.98 _____

8. 9.75 _____

9. 1,723.439 _____

Use with or after Lesson 10•4.

Practice Set 75

Write the *x*- and *y*-coordinates for each point.

1. A _____

2. B _____

3. C _____

4. D _____

5. E _____

6. F _____

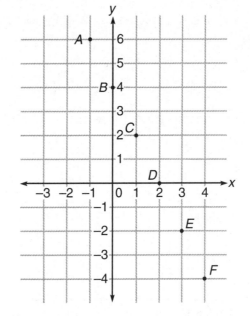

 Solve.

7. $\frac{1}{4}$ of 80 _____

8. $\frac{1}{6}$ of 240 _____

9. $\frac{4}{5}$ of 25 _____

10. $\frac{1}{40}$ of 1,200 _____

11. $\frac{1}{16}$ of 32 _____

12. $\frac{20}{30}$ of 90 _____

13. $\frac{5}{8}$ of 64 _____

14. $\frac{7}{9}$ of 81 _____

15. Lucky Video is going out of business. All videos are $\frac{1}{3}$ the regular price. Sal's Videos has a 50%-off sale. Which store is offering a better discount?

_____

16. If a video originally cost $15, how much will it cost at Lucky Video?

_____

Practice Set 75 continued

Circle the algebraic expression that best matches each situation.

17. Ellen jogs 3 miles each day. How many miles does
she jog in *d* days?

$3 * d$ $\qquad$ $3 + d$ $\qquad$ $d \div 3$

18. Tim takes $17 from his savings account. How much is
left in his savings account?

$s - 17$ $\qquad$ $s + 17$ $\qquad$ $17 * s$

19. Mr. Gomez drove *m* miles in 8 hours. How many miles
did he drive per hour?

$m * 8$ $\qquad$ $m / 8$ $\qquad$ $m + 8$

 Solve. Write the answer in simplest form.

20. $3\frac{1}{2} + 2\frac{1}{5} =$ _____

21. $9 - 5\frac{1}{8} =$ _____

22. $4\frac{1}{4} + 6\frac{1}{6} =$ _____

23. $8\frac{1}{3} * \frac{1}{2} =$ _____

24. $9\frac{1}{2} - 7\frac{2}{3} =$ _____

25. $4\frac{1}{3} * 2\frac{2}{5} =$ _____

Round 4,871,354 to the given place value.

26. million _____

27. ten-thousand _____

28. hundred-thousand _____

29. hundred _____

30. **Writing/Reasoning** Which of the answers to Problems 26–29
is closest to the original number? Explain how you know.

_____

_____

Practice Set 76

Match each of the events with one of the graphs.

Graph A

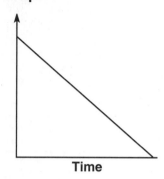

Time

Graph B

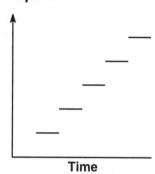

Time

Graph C

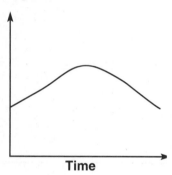

Time

Graph D

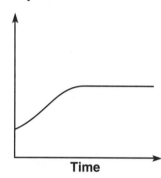

Time

1. The temperature outside on the patio during the day from sunrise to sunset:

_____

2. The amount of money in Fred's savings account; he makes a deposit at the end of each week:

_____

3. The volume of water in a sink as the water drains out:

_____

4. The temperature of a pan of water that heats on the stove to boiling:

_____

Practice Set **76** *continued*

Write the following numbers in digits.

5. eighteen trillion, six hundred twenty-seven billion, nine hundred million

_____

6. nine billion, one hundred fifty-six million

_____

7. six hundred fourteen trillion, four hundred million

_____

Write the following numbers in words.

8. 218,055,000 _____

9. 168,409,000 _____

10. 34,313,000,000 _____

11. 1,867,291,433 _____

_____

 Solve.

12. $65\overline{)910}$

13. $35\overline{)1{,}680}$

14.   7,940
      $+ \ 56{,}094$

15.   6,521
      $- \ 3{,}876$

16. $\dfrac{3}{4}$
     $-\dfrac{1}{4}$

17.   7,547
      $+ \ 4{,}546$

18. 364.0
      $- \ \ 4.5$

19.   3.970
      $- \ 1.050$

20. 18.2
      $* \ \ 4$

21. 363
      $- \ 34$

22. $\dfrac{1}{2} * 68$

23. 470
      $* \ 30$

Complete.

24. $10^4 =$ _____

25. $6^{\square} = 1{,}296$

26. $16^2 =$ _____

27. $9^{\square} = 729$

Practice Set 77

Find the area of each figure below.

1.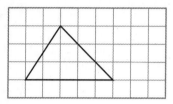

_____

2.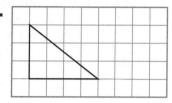

_____

3.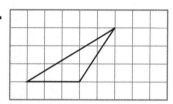

_____

4.

_____

5.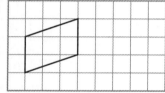

_____

6.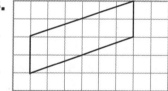

_____

Tell whether each number is *prime* or *composite*.

7. 51 _____   **8.** 323 _____   **9.** 119 _____

10. 101 _____   **11.** 99 _____   **12.** 67 _____

13. 2 _____   **14.** 143 _____   **15.** 891 _____

16. 77 _____   **17.** 210 _____   **18.** 417 _____

Practice Set 78

SRB
186–189
194

Diameter = 2 * Radius
Circumference = π * Diameter
Area = π * r²

1. Find the circumference.

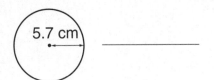

5.7 cm _____

2. What is the radius of the tire?

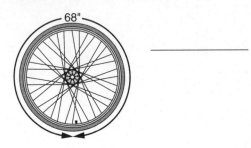

68" _____

Find the area of each circle.

3.

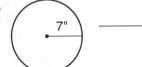

7" _____

4.

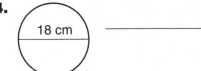

18 cm _____

Tell which measurement would be best for each situation.
Write *area*, *circumference*, or *perimeter*.

5. The amount of fence needed to enclose a rectangular garden.

_____

6. The amount of fabric needed to cover the top of a round table.

_____

7. The amount of lace needed to trim the outside edge of a circular table cloth.

_____

8. The amount of carpet needed to cover a rectangular floor.

_____

Use with or after Lesson 10·9.

Practice Set 78 continued

Complete each rate table below. Then answer the questions.

Example Susan's car gets about 35 miles per gallon of gasoline on the highway.

miles	35	70	105	140	175	210
gallons	1	2	3	4	5	6

9. How far can the car travel on 6 gallons of gas? _____

10. At 140 miles, how many gallons have been used? _____

The cows on the McCulhay farm each produce about 8 gallons of milk per day.

gallons	8						
cows	1	7	10	50	100	250	500

11. If 250 cows are milked every day, how many gallons of milk will be produced in a week?

_____

12. If 250 cows are milked for 30 days and 500 cows for 20 days, how many gallons of milk will be produced?

_____

Find the unit rate for each. Then circle the better buy.

13. 1 dozen eggs for $2.64 _____

1 half-dozen eggs for $1.50 _____

14. 3 cans of tomato sauce for $1.05 _____

2 cans of tomato sauce for $0.89 _____

15. 3 pounds of apples for $2.88 _____

40 pounds of apples for $26.00 _____

Practice Set 79

Write *prism, pyramid, cylinder, cone,* or *sphere.*

1. 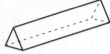 _____

2. _____

3. _____

4. _____

5. _____

6. _____

Write <, >, or = to compare the fractions. If necessary, rewrite the
fractions, using the least common denominator.

7. $\frac{1}{2}$ _____ $\frac{6}{7}$

8. $\frac{2}{3}$ _____ $\frac{8}{15}$

9. $\frac{11}{20}$ _____ $\frac{1}{12}$

10. $\frac{4}{9}$ _____ $\frac{7}{10}$

11. $\frac{5}{8}$ _____ $\frac{3}{5}$

12. $\frac{3}{21}$ _____ $\frac{2}{15}$

 Solve.

13. $2 \div \frac{3}{5} =$ _____

14. $\frac{2}{5} \div \frac{1}{10} =$ _____

15. $1\frac{1}{2} \div \frac{1}{4} =$ _____

16. $\frac{6}{5} \div \frac{2}{10} =$ _____

17. $2 \div \frac{1}{3} =$ _____

18. $3\frac{1}{5} \div \frac{2}{5} =$ _____

Use with or after Lesson 11·1.

Practice Set 80

Match each shape with at least one property. More than one property may apply.

1. triangular prism _____

2. tetrahedron _____

3. cube _____

4. cone _____

5. cylinder _____

A. At least 1 surface is a circle.

B. All surfaces are triangles.

C. All surfaces are the same shape.

D. The faces are rectangles, but not the bases.

E. There is only 1 base.

---

6. Write the missing numbers for the table.

Product	Exponential Notation	Standard Notation
8 * 8 * 8		
	10^5	
	$2^{\square}$	32
12 * 12 * 12		
		256

Complete the missing factors.

7. _____ * 70 = 350

8. 30 * _____ = 810

9. _____ * 11 = 6,600

10. 80 * _____ = 560

11. _____ * 9 = 360

12. _____ * 32 = 640

13. 11 * _____ = 770

14. 140 * _____ = 280

Practice Set 80 continued

Rename the following fractions as decimals.

15. $\frac{6}{10}$ _____

16. $\frac{3}{4}$ _____

17. $\frac{8}{24}$ _____

18. $\frac{16}{10}$ _____

19. $\frac{769}{1,000}$ _____

20. $\frac{28}{100}$ _____

21. $\frac{7}{8}$ _____

22. $2\frac{3}{4}$ _____

23. $\frac{18}{16}$ _____

24. $\frac{734}{100}$ _____

25. $5\frac{321}{1,000}$ _____

26. $\frac{180}{18}$ _____

Find the missing angle measurements without using a protractor.

27. $x =$ _____°

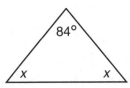

28. $a =$ _____°

29. $b =$ _____°

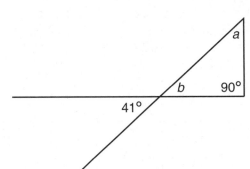

Write a number model that describes each shaded rectangle.

30.

_____

31.

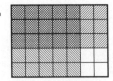

_____

32.

_____

Use with or after Lesson 11·2.

Practice Set 81

SRB
13–24
196–198

Find the volume of each figure. Include the units.

$$V = B * h$$

1.

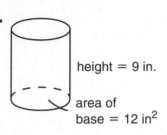

height = 9 in.

area of
base = 12 in²

_____

2.

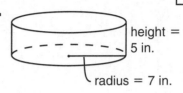

height =
5 in.

radius = 7 in.

_____

3.

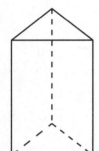

height = 12 in.

7 in.

← 6 in. →
base of prism

_____

COMPUTATION PRACTICE **Solve.**

4. 523
       * 20

5. 353
       − 75

6.   25
        − 7

7.   250
       + 2,700

8.   235
       + 2,352

9. 35)3,115

10. 542
        * 21

11. 26)1,066

12. 7,345
        − 3,462

13.   691
       + 5,137

14. 4,807
        − 929

15. 17)6,243

Practice Set 81 *continued*

Complete the "What's My Rule?" tables and rule boxes.

16.

Rule	in	out
out = in * 300	9	
	12	
	15	
	25	
	100	

17.

Rule	in	out
	7	17.5
	10	25
		12.5
	18	
	100	250

18.

Rule	in	out
out = in − 55	80	
	160	
		90
	2,400	
		1,200

19.

Rule	in	out
	800	20
	160	
		90
	2,400	
	4,800	120

Tamara wants to save $80 during her summer vacation.
After 2 weeks she has saved $16.

20. What fraction of the $80 did she save in the first 2 weeks? _____

21. What percent did she save? _____

22. If Tamara continues saving at the same rate, how

much will she save after 5 weeks? _____

23. At this rate, how long will it take her to save $80? _____

24. **Writing/Reasoning** Explain how you found the answer to
Problem 23.

_____

_____

_____

Use with or after Lesson 11·3.

Practice Set 82

Find the volume of each figure. Include the units.

> Prism and Cylinder $V = B * h$
> Pyramid and Cone $V = \frac{1}{3} * (B * h)$

1.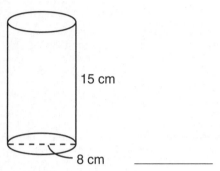

15 cm

8 cm _____

2. 8 cm

15 cm _____

3.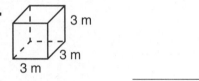

3 m

3 m

3 m _____

4. 3 m

3 m

3 m _____

Write each fraction as a mixed number or a whole number.

5. $\frac{30}{12}$ _____

6. $\frac{24}{8}$ _____

7. $\frac{9}{4}$ _____

8. $\frac{50}{15}$ _____

9. $\frac{85}{20}$ _____

10. $\frac{66}{9}$ _____

 Solve.

11. $-20 - (-56) =$ _____

12. $-4 + (-24) =$ _____

13. $17 + (-20) =$ _____

14. $-3 - 42 =$ _____

15. **Writing/Reasoning** When adding a positive and a negative number, how can you tell by looking at the problem whether your answer will be positive or negative?

_____

_____

Practice Set 82 continued

Complete. You may use a calculator.

16. $8^3 =$ _____

17. $7^4 =$ _____

18. $3^6 =$ _____

19. $4^5 =$ _____

20. $6^7 =$ _____

21. _____$^3 = 125$

22. $8^{\square} = 32,768$

23. _____$^6 = 64$

Find the total cost.

24. 7 rulers that cost 45¢ each _____

25. 9 scissors that cost $2.65 each _____

26. 28 books that cost $1.15 each _____

 Solve.

27. $\begin{array}{r} 9,376 \\ + 4,329 \\ \hline \end{array}$

28. $\begin{array}{r} 1,754 \\ + 2,845 \\ \hline \end{array}$

29. $\begin{array}{r} 1,000 \\ - 43 \\ \hline \end{array}$

30. $\begin{array}{r} 20,000 \\ * 5 \\ \hline \end{array}$

31. $300 * \frac{1}{3}$

32. $\begin{array}{r} 5\frac{1}{4} \\ + 3\frac{1}{2} \\ \hline \end{array}$

33. $\begin{array}{r} 532 \\ - 58 \\ \hline \end{array}$

34. $\begin{array}{r} 234 \\ - 18 \\ \hline \end{array}$

35. $\begin{array}{r} 234 \\ * 52 \\ \hline \end{array}$

36. $\begin{array}{r} 721 \\ * 82 \\ \hline \end{array}$

37. $\begin{array}{r} 687 \\ * 25 \\ \hline \end{array}$

38. $\begin{array}{r} 9\frac{3}{4} \\ + 10\frac{1}{4} \\ \hline \end{array}$

Use with or after Lesson 11·4.

Practice Set 83

Complete.

1. 2 pints = _____ quart(s)

2. 8 ounces = _____ cup(s)

3. 18 quarts = _____ cup(s)

4. 1 gallon = _____ pint(s)

5. 3 half-gallons = _____ ounce(s)

6. 7 pints = _____ quart(s)

7. 100 ounces = _____ cup(s)

8. 4 pints = _____ cup(s)

9. Ava is making lemonade for 32 people. She wants to make enough lemonade for each person to have two 8 oz glasses. How many gallons of lemonade should she make? _____

Measure each angle to the nearest degree.

10.

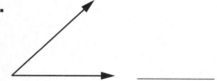 _____

11.

 _____

12.

 _____

13.

 _____

14.

_____

15.

 _____

 Solve.

16. $\frac{4}{5} - \frac{2}{3} =$ _____

17. $\frac{8}{9} - \frac{1}{2} =$ _____

18. $2\frac{4}{5} + \frac{11}{5} =$ _____

19. $\frac{12}{5} - 2 =$ _____

20. $\frac{12}{16} - \frac{5}{16} =$ _____

21. $\frac{34}{20} + \frac{2}{5} =$ _____

22. $\frac{3}{9} - \left(-\frac{1}{4}\right) =$ _____

23. $9,000 - \frac{15}{16} =$ _____

24. $\frac{5}{6} - \left(-\frac{1}{4}\right) =$ _____

Practice Set 83 *continued*

Write the number pairs for each point.

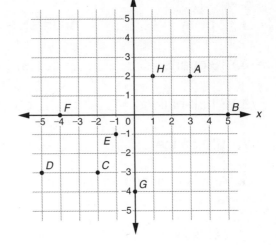

25. *A* (_____, _____)

26. *C* (_____, _____)

27. *E* (_____, _____)

28. *G* (_____, _____)

Name the point at each number pair.

29. (−5, −3) _____

30. (1, 2) _____

31. (5, 0) _____

32. (−4, 0) _____

Choose the graph that best represents the event described.

33. Joey fills his cup with water. He adds an ice cube. He adds a second ice cube. He takes a sip. He takes another sip. He drinks the rest of the water in one gulp.

_____

A

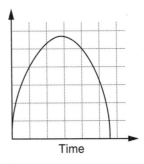

Time

34. A football is thrown straight up into the air.

_____

B

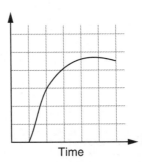

Time

35. A frozen soft pretzel is removed from the freezer. It is heated in the oven. Then it is placed on a plate.

_____

C

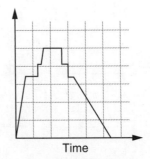

Time

Use with or after Lesson 11•6.

Practice Set 84

Find the volume (*V*) and the surface area (*S*) for each figure.

Shape	Volume	Surface Area
Rectangular Prism	$V = B * h$	$S = 2 * ((l * w) + (l * h) + (w * h))$
Cylinder	$V = B * h$	$S = (2 * \pi * r^2) + ((2 * \pi * r) * h)$
Square Pyramid	$V = \frac{1}{3} * (B * h)$	$S = 4 * (\frac{1}{2} * b * H) + s^2$

1.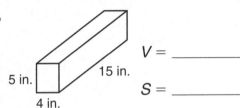

5 in.   15 in.   4 in.

V = _____

S = _____

2.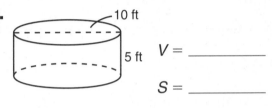

10 ft   5 ft

V = _____

S = _____

3.

4 m   5 m   6 m   6 m

V = _____

S = _____

4. **Writing/Reasoning** Explain how you found the surface area of the cylinder.

_____

_____

_____

Make a magnitude estimate for the quotient. Is the solution in the *tenths, ones, tens,* or *hundreds*?

5. 8.2 / 12 _____

6. 52.3 / 9 _____

7. 956 / 3 _____

8. 591 / 25 _____

9. 1,226 / 85 _____

10. 12.8 / 9 _____

11. 386 / 19 _____

12. 5.73 / 8 _____

Practice Set 84 continued

Ms. Lewis supervises newspaper delivery routes. She made
a stem-and-leaf plot to show the number of papers delivered
on each paper route.

13. How many paper routes does Ms. Lewis supervise? _____

14. What is the maximum? _____

15. What is the minimum? _____

16. What is the mode? _____

17. What is the median? _____

18. What is the mean? _____

**Number of Newspapers
Delivered on Each Route**

Stems (10s)	Leaves (1s)
2	2 5 5 9
3	1 6 6 8
4	0 3 3 3 5
5	2 3 7
6	2

Add parentheses to the number sentences to make them correct.

19. $8 * 10 - 6 = 32$

20. $63 = 5 + 2 * 9$

21. $38 - 32 - 24 = 30$

22. $27 - 28 + 37 = -38$

23. $6 * 4 + 4 * 12 = 576$

24. $517 = 11 * 9 + 38$

25. $9 * 3 + 9 * 6 = 81$

26. $99 = 6 * 24 + 35 - 80$

27. **Writing/Reasoning** Write and solve an equation that has
parentheses. Then show how you can get a different answer
by moving the parentheses. Explain why the two answers are
different.

_____

_____

_____

Use with or after Lesson 11·7.

Practice Set 85

List all of the factors of each number. Then circle the greatest common factor of each pair of numbers.

1. 12 _____   20 _____

2. 30 _____   36 _____

3. 8 _____   52 _____

4. 18 _____   60 _____

Write the prime factorization of each number.

5. 15 _____   **6.** 28 _____   **7.** 35 _____

8. 40 _____   **9.** 42 _____   **10.** 48 _____

11. Complete the table.

Product	Exponential Notation	Standard Notation
	10^8	
$6 * 6 * 6 * 6 * 6$		
	$4^{\square}$	1,024
	$\underline{}^6$	15,625
$9 * 9$		
	3^4	
	$7^{\square}$	343
	$\underline{}^5$	32,768

Practice Set 86

Find all of the possible choices in each situation. You may want to make a tree diagram to help you list the choices.

1. Ellen packed her suitcase for the weekend. She packed one pair of shorts, one pair of pants, and three tops: pink, striped, and blue. How many outfits can she make?

2. A sack lunch from the snack bar has a beverage, a sandwich, and a dessert. The beverages are milk, apple juice, and orange juice. The sandwiches are cheese, tuna fish, and ham. The desserts are apple pie and cookies. How many different sack lunches are available?

The spinner is equally likely to stop in each section. Find the probability that the spinner will stop on . . .

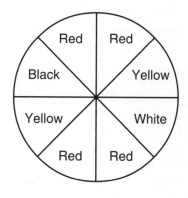

3. yellow _____

4. red _____

5. white _____

6. blue _____

7. Which two colors have the same probability that the spinner will stop on them?

_____

Find the least common multiple for each pair of numbers.

8. 3, 12 _____

9. 6, 8 _____

10. 9, 10 _____

11. 12, 16 _____

12. 8, 12 _____

13. 5, 12 _____

14. 17, 3 _____

15. 2, 7 _____

16. 3, 4 _____

Use with or after Lesson 12·2.

Practice Set 86 *continued*

Complete the "What's My Rule?" tables.

17.

Rule		in	out
out = in + 116		165	
		−433	
			97
		114	
			−82

18.

Rule		in	out
out = in * 540		21	
		43	
			4,320
		54	
		109	

Find the perimeter of the figures below. Include the units.

19.

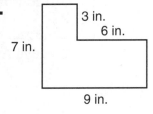

6 cm
9 cm
5 cm
17 cm

_____

20.

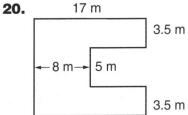

17 m
3.5 m
8 m
5 m
3.5 m

_____

21.

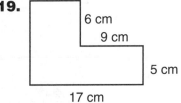

3 in.
6 in.
7 in.
9 in.

_____

22.

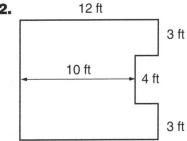

12 ft
3 ft
10 ft
4 ft
3 ft

_____

23. **Writing/Reasoning** Explain how you found the answer to Problem 20.

_____

_____

_____

Practice Set 87

 Find the percents.

1. 34% of 500 _____

2. 250% of 60 _____

3. 10% of 350 _____

4. 80% of 600 _____

5. 1% of 54 _____

6. 10% of 5.77 _____

7. 15% of 20 _____

8. 20% of 540 _____

Solve each number sentence by finding the value of the variable.

9. $a = (2 * 18) / 3$ _____

10. $6.4 / b = 3.2 / 2$ _____

11. $c = (2.8 + 9) / 2$ _____

12. $65 + (13 * 11) = d$ _____

13. $(8.1 + 2.9) / 2 = b$ _____

14. $c = (6.2 + 3\frac{4}{5}) / 5 =$ _____

Ms. Jewel's class collected the following information about favorite clubs in their school. They interviewed 200 students.

15. How many students liked band best?

_____

16. How many more students liked community service than ecology? _____

17. How many students liked creative arts and journalism?

_____

24%
Community Service

36%
Band

12%
Journalism

12%
Creative Arts

16%
Ecology

18. **Writing/Reasoning** The school has received additional money to fund one club. The community service club and the creative arts club are already fully funded. Which one of the remaining three clubs do you think should receive the money? Why?

_____

_____

154

Practice Set 88

In a school election, 210 votes were cast. Two-fifths of the students voted for Lisa, and $\frac{1}{5}$ of the students voted for Christy. The rest of the votes were for write-in candidates.

1. How many votes did Christy get? _____

2. How many votes did Lisa get? _____

3. How many votes were for write-in candidates? _____

210 votes

Lisa   Christy

Complete the rate table. Then answer the questions.

Sparkling Cleaners can dry-clean 40 shirts per hour.

shirts	80					
hours	2	4	6	8	10	12

4. How many shirts can Sparkling Cleaners clean in an 8-hour work day?

_____

5. How many shirts can it clean in a 40-hour work week?

_____

6. If the dry cleaners hired 3 shifts of workers so that it could stay open 24 hours a day, how many shirts could it clean in 1 day?

_____

Write each number in scientific notation.

7. 6,000 _____

8. 3 million _____

9. 50 thousand _____

10. 70 billion _____

11. 4 hundred thousand _____

12. 4 trillion _____

13. 512 thousand _____

14. 36 million _____

Practice Set 88 continued

SRB
243
228 229

Solve.

15. A canister shaped like a cylinder has a radius of 3 inches and a height of 10 inches. What is the volume of the canister?

_____

16. A round rug has a diameter of 30 inches. What is the circumference?

_____

17. A punch bowl holds 3 gallons. How many pint containers of juice will fill the punch bowl?

_____

18. The weight of an object on the Moon is $\frac{1}{6}$ the weight of the same object on Earth. If an object weighs $24\frac{1}{2}$ pounds on the Moon, how many pounds does it weigh on Earth?

_____

19. There is a 60% chance of rain for tomorrow. What is the chance that it will not rain?

_____

20. **Writing/Reasoning** In Problem 19, would you say that the chance of rain for tomorrow is *likely*, *unlikely*, or *equally likely*? Explain your answer.

_____

_____

Solve the pan-balance problems.

21. ($\frac{1}{2}$ cantaloupe)

Three cantaloupes weigh as much as _____ apples.

22.

000 10 ☐

One cube weighs as much as _____ marbles.

Use with or after Lesson 12•4.

Practice Set 89

Write a number model for each problem. Then solve.

1. There are 270 students at Carter Elementary School.
   Two out of three students buy lunch in the cafeteria.
   How many students buy lunch?

   _____

2. Marilyn had 96 postcards in her collection.
   One in four are from foreign countries.
   How many of the postcards are from foreign countries?

   _____

Solve.

3. The diameter of Earth is about 4 times the diameter
   of the Moon. If Earth's diameter is about 8,000 miles,
   what is the approximate diameter of the Moon?

   _____

4. What is the circumference of Earth? _____

5. What is the circumference of the Moon? _____

6. About how many times larger is the circumference
   of Earth than the circumference of the Moon?

   _____

7. Ms. Kennedy presents a program about wild animals,
   using 3 kinds of animals. She has 3 birds, 4 lizards,
   and 2 squirrels. She uses 1 of each type of animal each
   time. What are the total number of different ways she can
   choose the animals for her shows?

   _____

8. An ice cream shop offers 3 sizes of ice cream cones
   and 16 different ice cream flavors. How many different
   ways are there to order an ice cream cone?

   _____

Practice Set 90

SRB
68–70,
233 234

1. Marco runs at a rate of 10 minutes per mile. Complete the table.

Minutes	10	20	45		95
Miles	1			7.5	

2. Graph the data from the table and connect the points.

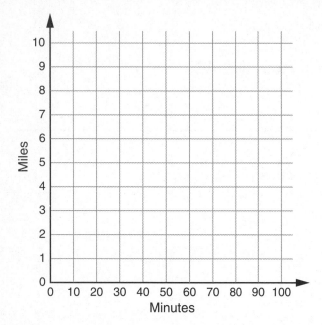

3. **Writing/Reasoning** If Marco continued running at this rate, how far could he run in 150 minutes? Explain your answer.

_____

_____

_____

COMPUTATION PRACTICE **Add.**

4. $1\frac{2}{3} + 4 =$ _____

5. $\frac{4}{5} + 3\frac{1}{2} =$ _____

6. $\frac{3}{8} + \frac{2}{3} =$ _____

7. $3 + 5\frac{7}{8} =$ _____

8. $6\frac{1}{4} + 9 =$ _____

9. $\frac{5}{6} + 7\frac{1}{2} =$ _____

Use with or after Lesson 12•7.

Test Practice ⟨1⟩

Fill in the circle next to your answer.

1. This factor tree shows how to write 20 as the product of prime numbers. Which of the following shows 72 as a product of prime numbers?

   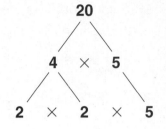

   Ⓐ 6×12

   Ⓑ $2 \times 3 \times 3$

   Ⓒ $2 \times 2 \times 2 \times 9$

   Ⓓ $2 \times 2 \times 2 \times 3 \times 3$

2. The first five prime numbers are 2, 3, 5, 7, and 11. What are the prime numbers between 50 and 60?

   Ⓐ 53, 59

   Ⓑ 52, 54, 56, 58

   Ⓒ 51, 53, 55, 57, 59

   Ⓓ 51, 52, 54, 55, 56, 57, 58

3. Tara's fish tank has a capacity of 100 liters. Jamel's fish tank has a capacity of 75.5 liters. How many more liters does Tara's fish tank hold?

   Ⓐ 24.5 liters

   Ⓑ 25 liters

   Ⓒ 25.5 liters

   Ⓓ 26 liters

4. Manny's little sister made the following designs with her blocks. Which is an example of a **tessellation**?

   Ⓐ 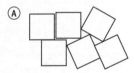   Ⓑ    Ⓒ 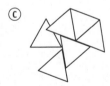   Ⓓ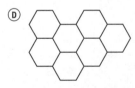

5. An acute angle has a measure that is

   Ⓐ greater than 180°.

   Ⓑ less than 180 but greater than 90°.

   Ⓒ greater than 90°.

   Ⓓ less than 90°.

Test Practice 1 continued

Fill in the circle next to your answer.

6. Which would be the **best** tool to measure angles of shapes on a graph?

 (A)

tape measure

(B) compass

(C)

balance scale

(D) 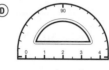 protractor

7. Ciro left a paper clip on his desk in this position.

When he got back to his desk, he noticed that someone had rotated the clip 180°. Which of the following shows the new position of the paper clip?

(A)

(B)

(C)

(D)

8. Which is the best magnitude estimate for 57 * 412?

(A) 50 * 400

(B) 50 * 450

(C) 60 * 400

(D) 60 * 450

Use with or after Unit 3.

Test Practice 2

Fill in the circle next to your answer.

1. This number line shows the distance in miles of some buildings from Corinne's house.

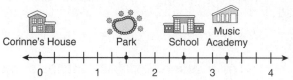

   How far from Corinne's house is the Music Academy?

   Ⓐ $2\frac{1}{2}$ miles   Ⓑ $3\frac{1}{4}$ miles   Ⓒ 3.1 miles   Ⓓ 3.5 miles

2. The data below show the results of a survey of students' favorite beverages.

   Milk: 80 votes   Apple Juice: 65 votes

   Orange Juice: 72 votes   Spring Water: 57 votes

   Diego wants to display the results in a bar graph. He is thinking of using the intervals 1, 2, 10, or 100 for the vertical axis. Which would be the **best** choice?

   Ⓐ 1   Ⓑ 2   Ⓒ 10   Ⓓ 100

3. Carmela, Darin, Eldon, and Amelia have a math test today. The hours they studied last night are shown in the table to the right.
   Which statement correctly compares the amount of time that Darin and Amelia studied?

   Ⓐ $\frac{1}{2} > 0.60$ Ⓑ $\frac{1}{2} = 0.60$

   Ⓒ $0.60 > \frac{1}{2}$ Ⓓ $0.60 < \frac{1}{2}$

Hours Of Study	
Student	**Hours of Study**
Carmela	0.75
Darin	$\frac{1}{2}$
Eldon	2.20
Amelia	0.60

4. This circle graph shows how Dolores spends her spare time. What percent of her spare time did she spend reading?

   Ⓐ 20%   Ⓑ 25%

   Ⓒ 30%   Ⓓ 35%

HOW DOLORES SPENDS HER SPARE TIME

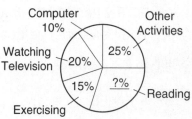

Test Practice 2 continued

5. This line graph shows the average rainfall in Tampa, Florida, for each month.

Which conclusion can you draw from this graph?

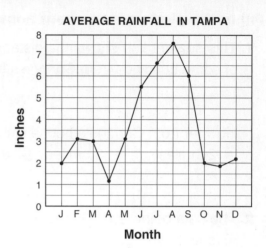

AVERAGE RAINFALL IN TAMPA

Ⓐ The average rainfall is 5 inches or more for 6 months.

Ⓑ The average rainfall for February and March is about the same.

Ⓒ The average rainfall for June is less than the average rainfall for October.

Ⓓ Tampa gets more rain in the winter months than in the summer months.

6. Tanisha is shopping for a computer game for her little brother. She researched the price at seven different stores.

COMPUTER GAME PRICES

Stem	Leaves
1	9 9
2	2 3 4 6
3	0

Key: 1|9 means $19

What is the median price of the computer games she researched?

Ⓐ $19   Ⓑ $23   Ⓒ $26   Ⓓ $30

7. How many minutes equal $\frac{1}{6}$ of an hour?

Ⓐ 20 minutes   Ⓑ 15 minutes

Ⓒ 10 minutes   Ⓓ 5 minutes

8. It was 55°F when Mark left for school. By lunchtime it was 67°F. Which number sentence best models this situation?

Ⓐ $55 + t = 67$ Ⓑ $55 + 67 = t$

Ⓒ $67 + t = 55$ Ⓓ $67 + t = 67$

Test Practice 3

Fill in the circle next to your answer.

1. Ms. Velez charged $25 per hour for 3 hours of work and $40 for parts for a repair job. Which expression could be used to find the total amount she charged?

   Ⓐ $(25 \times 3) + 40$
   Ⓑ $(25 + 3) + 40$
   Ⓒ $(25 \div 3) + 40$
   Ⓓ $(25 + 3) - 40$

2. $-3 + (+8) =$ _____

   Ⓐ 11
   Ⓑ 5
   Ⓒ -5
   Ⓓ -11

3. There were $1\frac{1}{8}$ pounds of potato salad and $1\frac{3}{4}$ pounds of pasta salad left over after Calvin's dinner party. About how many pounds of salad were left over?

   Ⓐ 2
   Ⓑ $2\frac{1}{2}$
   Ⓒ 3
   Ⓓ $3\frac{1}{2}$

4. Benjamin has a muffin recipe that calls for $\frac{1}{2}$ cup of apple juice concentrate. He wants to make just $\frac{1}{2}$ of the recipe. About how much apple juice concentrate should he use?

   Ⓐ $\frac{1}{4}$ cup
   Ⓑ $\frac{1}{3}$ cup
   Ⓒ $\frac{3}{4}$ cup
   Ⓓ $1\frac{1}{12}$ cups

5. This picture shows the first four figures in a pattern.

   How many blocks will be in the sixth figure of this pattern?

   Ⓐ 36   Ⓑ 49   Ⓒ 50   Ⓓ 54

Test Practice 3 continued

6. The arrows show the path Miranda takes to get to Linda's house. What are the coordinates of Linda's house?

Ⓐ (5,2)   Ⓑ (5,3)

Ⓒ (3,5)   Ⓓ (2,5)

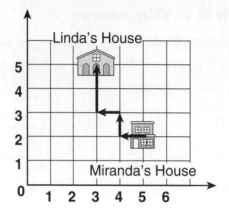

7. Leon's family is packing food in cartons to take to a family reunion. How many cubic inches of food can this carton hold?

Volume = length × width × height

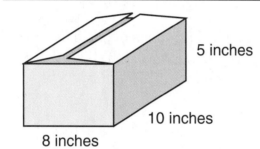

5 inches

10 inches

8 inches

Ⓐ 400 in.³   Ⓑ 40 in.³   Ⓒ 400 in.²   Ⓓ 40 in.²

8. What is the formula for the area of a triangle?

Ⓐ $\frac{1}{2} b + h$ Ⓑ $\frac{1}{2} b * h$ Ⓒ $b * h$ Ⓓ $\frac{1}{2} b \div h$

9. If the height of the larger triangle is CD, which side of the triangle is the base?

Ⓐ $\overline{AC}$ Ⓑ $\overline{BC}$ Ⓒ $\overline{AB}$ Ⓓ $\overline{CB}$

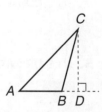

164

Test Practice ◆ 4

Fill in the circle next to your answer.

1. Carol graphed these points as part of a pattern.

If the pattern continues, what will be the missing value in this ordered pair: (10, __)?

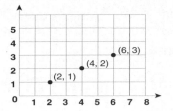

Ⓐ 4   Ⓑ 5

Ⓒ 6   Ⓓ 8

2. This table shows a set of input and output numbers for a function.

In	1	2	3	4	5	6
Out	2	8	18	32	?	?

What are the next two output numbers?

Ⓐ 40 and 52   Ⓑ 50 and 68   Ⓒ 50 and 72   Ⓓ 64 and 128

3. Estimate the area of the circle.

Ⓐ 12.5 cm²   Ⓑ 16 cm²

Ⓒ 36 cm²   Ⓓ 50.2 cm²

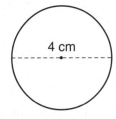

4 cm

4. Ivana is playing a board game, using this spinner.

What is the probability that she will move ahead 3 spaces after her **next** spin?

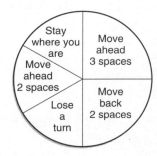

Ⓐ $\frac{1}{2}$ Ⓑ $\frac{1}{3}$

Ⓒ $\frac{1}{4}$ Ⓓ $\frac{1}{6}$

Test Practice ◆ 4 ◆ *continued*

Fill in the circle next to your answer.

5. What is the measure of ∠C?

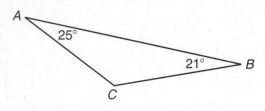

- Ⓐ 44°
- Ⓑ 46°
- Ⓒ 134°
- Ⓓ 294°

6. Mrs. James works for a company that makes calculators. Each calculator comes in a box that measures 12 inches by 6 inches by 2 inches. How many boxes will she be able to pack in a shipping crate that measures 24 inches by 24 inches by 12 inches?

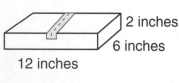

- Ⓐ 6,912
- Ⓑ 2,304
- Ⓒ 144
- Ⓓ 48

7. At Broadmoor Elementary School, 2 out of 5 students play a sport. There are 225 students at the school. How many students play a sport?

- Ⓐ 45
- Ⓑ 80
- Ⓒ 90
- Ⓓ 100

8. There are 20 marbles in a complete set. How many marbles are in $\frac{4}{5}$ of a set?

- Ⓐ 4
- Ⓑ 5
- Ⓒ 12
- Ⓓ 16

Use with or after Unit 10.

Name Courtney Tran ♡

WrightGroup.com

The **McGraw-Hill** Companies

ISBN 978007622505-7
MHID 007622505-4

90000

Mc
Graw
Hill **Wright Group**

4 Fold the circle in half. Fold it in half again. Fold again as you would to cut a paper snowflake.

5 Cut pieces out of the folded paper to make a design.

6 Open the cut paper circle and center it on the rim cut from the margarine tub. Snap the lid rim over the paper and the tub rim to hold the paper in place.

Hang your cut paper design in a sunny window for the light to shine through. Make three or four cut paper designs in different colors and hang them at different heights in the window for a very pretty Shavuot display.

ABOUT THE AUTHOR AND ARTIST

Twenty years as a teacher and director of nursery school programs in Oneida, New York, have given Kathy Ross extensive experience in guiding children through craft projects. She collects teddy bears and paper dolls, and her craft projects have frequently appeared in *Highlights* magazine. She is the author of the Holiday Crafts for Kids series, including *Crafts for Hanukkah* and *Every Day is Earth Day*. Her other authorial credits include *Gifts to Make for Your Favorite Grownup*, *Crafts for Kids Who Are Wild About Dinosaurs*, and *Crafts for Kids Who Are Wild About Outer Space*.

A native of Buffalo, New York, Melinda Levine is a graduate of the University of Buffalo with a degree in English Literature. She has illustrated a number of children's books, among them *Where's Cupcake?* and *I Am*, both by Catherine Peters, *Fun With Hats* by Lucy Malka, and *How to Get to Harry's House* by Merry Banks. Her artwork also appears in magazines. She is the author of a book on living rooms in the Color in Living Series published by Rockport Press. The artist currently lives in Oakland, California, with her husband and teenage daughter.